STUDENT'S SOLUTIONS MANUAL

DENISE MEWBORN
University of Georgia

MATHEMATICS FOR ELEMENTARY SCHOOL TEACHERS
FOURTH EDITION

Phares O'Daffer
Illinois State University

Randall Charles
San Jose State University

Thomas Cooney
University of Georgia

John A. Dossey
Illinois State University

Jane Schielack
Texas A&M University

PEARSON
Addison
Wesley

Boston San Francisco New York
London Toronto Sydney Tokyo Singapore Madrid
Mexico City Munich Paris Cape Town Hong Kong Montreal

Reproduced by Pearson Addison-Wesley from electronic files supplied by the author.

Copyright © 2008 Pearson Education, Inc.
Publishing as Pearson Addison-Wesley, 75 Arlington Street, Boston, MA 02116.

ISBN - 13:   978-0-321-44858-3
ISBN - 10:   0-321-44858-8

1 2 3 4 5 6 BB 09 08 07 06

SOLUTIONS

SECTION 1.1

1. Responses will vary. They may include technical geometric terms such as perpendicular, parallel, right angle or more generic terms such as 'the same' and 'across and down'. A good way to asses the quality of the directions is to give them to someone else and see what figure is produced by following the directions.

3. Responses will vary among students. Some may choose to use words describing sets of numbers: rational, real, etc. Others may choose to use geometric terms. And yet other students may choose to use terms associated with computations.

5. Responses will vary among students. Possibilities include: **XXX** XX; 0.6, 6/10.

7. Responses will vary. Possibilities include:
    Verbal: Michael has 4 children, and he wants to share his estate equally among them. What portion if his estate will each one receive?
    Visual:

    Numerical: 1/4 or 0.25…
    Graphic:

    0   1

9.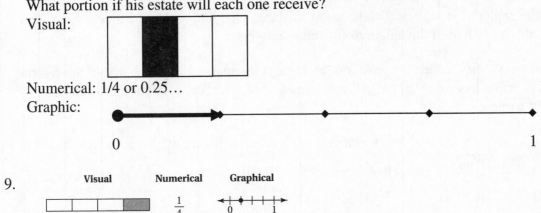

11. Responses will vary. One possibility is: Cut the pizza in half along a diameter. Cut the pizza in half along a diameter again, making sure the two cuts are perpendicular to one another. Take 1 of the pieces of pizza that has been created.

13. Responses will vary among students. The message conveyed by the graph is ambiguous. Although better than 1 in 10 women do not conduct self-exams, over half (38% + 4% + 11%) do follow the once per month recommendation. Are we pleased that over half of the women do work for early detection and thus enhance survival chances or are we displeased that a significant proportion do not follow the recommendation?

15. **Fifty three percent** (38% + 4% + 11%) of the women responding to the survey perform self-exams at least once a month.

17. a.   The value in B2 is 10% of the value in A2. So **B2 = .10 * A2**.
    b.   C2 can be directly calculated from A2 (**C2 = .90 * A2**) or by subtracting B2 from A2.

19. A **programable scientific** calculator would be appropriate. Another approach would be to use a graphing calculator to graph the equation and then use the trace function.

21. Discussion of the usefulness of the software will vary with the experiences and perspectives of the students. One important characteristic of GES is that it permits many trials of a situation rapidly. Thus conjectures may be formulated or counter-examples found. The software enables students to obtain tables of measures easily and rapidly. GES emphasizes the inductive component of mathematics in non-arithmetic areas.

23. Responses will vary. Possibilities include:
    a. What is the most common response of families to migraine sufferers?
    b. The graph does not convey characteristics of migraine sufferers themselves nor the number of sufferers.
    c. This is a matter of opinion. Migraine sufferers have to cope with the headaches, families have to cope with persons with migraines.
    d. Families may have placed themselves in more than one category.

25. The results of the exploration suggest that the segment joining the midpoints of 2 sides of a triangle has a length equal to one half of the length of the remaining side.

27. All of the figures in the first collection have a right angle at the marked vertex. None of the figures in the second collection has a right angle at the marked vertex. The first collection of figures illustrates "right angleness."

29. The estimates will vary among students. Geometry exploration software produces the result that EFGH is one fifth of ABCD.

31.

Hens	Rabbits	Feet
50	0	50
40	10	120
30	20	140

Thus the farmer has **30 hens and 20 rabbits**. Note that in employing a spreadsheet it is often useful to begin with rather coarse increments to establish the range containing the required value. One may then use smaller increments. In this case the broader increment contained the exact answer.

33. Thales discovered that angles inscribed in semicircles are right angles.

SECTION 1.2

1. The four examples show that adding 0 to a number results in the number. One might generalize from these examples that the sum of 0 and any number is the original number.

3.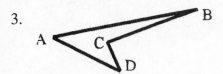

ABCD is not a square.

4.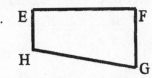

EFGH is not a rectangle.

5. $20 \div 4 = 5$; 5 is not an even number.

7. 2, 6, 10, 14, 18, **22, 26, 30: arithmetic ($d = 4$).**

9. 1, 4, 9, 16, 25, **36, 49, 64: neither** (sequence is squares of consecutive integers).

11. 1, 3, 7, 15, 31, **63, 127, 255: neither** (to produce next term double the current term and add 1).

13. 1, 2, 4, 7, **11, 16, 22: neither** (the differences between consecutive terms are the consecutive natural numbers).

15. 25, 21, 17, 13, **9, 5, 1: arithmetic, ($d = -4$).**

17. Responses will vary. They may include the ideas that:
    a. Conclusions are arrived at through inductive reasoning by assuming that some common aspect of several specific instances applies to the class of similar cases. For example, because GES shows that the sum of the angles of each of several triangles sum to 180°, we conclude that the angles of all triangles sum to 180°. The conclusion could be shown to be false by a *single specific* counterexample.
    b. Conclusions arrived at deductively begin with some general statement accepted as true either by agreement or by prior deductive proof (obviously the entire process of deduction rests upon general statements true by agreement). These statements include requirements that, if met by a more specific situation, allow the conclusion of the general truth to be applied to the more specific situation. For example, we agree that if the sun shines tomorrow then the game will be played. Upon arising we agree that the sun is shining. We can conclude that the game will be played. A conclusion based on deductive reasoning may be challenged by attacking the *form* of the argument rather than the specific content.

19. Every square is not a rectangle. The negation is false.

21. Hypothesis: an even number is multiplied by an odd number; conclusion: the product is an even number.

23. Hypothesis: a product is in high demand; conclusion: the price will increase.

25. **Inductive reasoning** based on the calculator display suggests the numbers form a geometric sequence with r = 1/2. Therefore, continuing the pattern based on this conclusion, a prediction of 1 is reasonable.

27. Recall that p→q is false only when p is true and q is false.

p	q	p→q
T	T	T
T	F	F
F	T	T
F	F	T

  The statement is false if Serena wins the second set and loses the match.

29. A right triangle has one right angle and two acute angles. The conjunction is true because both the original statements are true.

31. The number 9 is either prime or odd. The disjunction is true because the second statement is true.

33. If a triangle has three congruent sides, then it has three congruent angles. If a triangle has three congruent angles, then it has three congruent sides.

35 and 37 are examples of **denying the conclusion.**

39 is an example of **affirming the hypothesis**.

41. This is an example of **deductive** reasoning.

43. This is an example of **inductive** reasoning.

45. This reasoning is **invalid.** It is an example of **assuming the converse.**

47. This example is **invalid.** It is an example of **assuming the inverse**.

49. If a number is not positive, then its square is not positive. (False—the number could be negative.)

51. (a) If a figure is a rhombus, then it has four congruent sides. (b) If a figure has four congruent sides, then it is a rhombus.

53. No, it could be a non-square rectangle.

55. The conjunction is false because $p \wedge \sim q$ can only be true if both p and $\sim q$ are true. But we know that q is true, and it is not possible for q and $\sim q$ to be true simultaneously.

57. The biconditional statement will be false. I order for $p \rightarrow q$ to be true, both $p \rightarrow q$ and $q \rightarrow p$ must be true.

59.

p	q	$\sim p$	$\sim q$	$\sim p \rightarrow q$	$\sim q \rightarrow p$
T	T	F	F	T	T
T	F	F	T	T	T
F	T	T	F	T	T
F	F	T	T	F	F

61. The product in the first column is 1 more than the product in the second column. Then in the next pair, the product in the first column is 1 less than the product in the second column. This pattern continues with other numbers in the sequence:

1 x 3	3	1 x 2	2
1 x 5	5	2 x 3	6
2 x 8	16	3 x 5	15
3 x 13	39	5 x 8	40
5 x 21	105	8 x 13	104
8 x 34	272	13 x 21	273
13 x 55	715	21 x 34	714

63. Responses will vary. Possibilities include: Assume that the conditional statement "If a quadrilateral is a rhombus, then the sides of that quadrilateral are all the same length." is true (It is.). Now, let geometric figure ABCD be a rhombus. By affirming the hypothesis, we can conclude the sides of ABCD are the same length.

65. Responses will vary. A possibility is the set: $3^2 = 9$, $17^2 = 289$, $21^2 = 441$.

67. a. The conclusion is that figure ABCD is a rectangle. We have affirmed the hypothesis.
    b. The conclusion is that figure ABCD is a polygon, again by affirming the hypothesis.
    c. The conclusion is that 15 is an odd number by the same reasoning as in a and b.

69. The inverse statement.

71. **The generalization follows** from the examples given. However, in exercise 71 the statement can be shown to be **incorrect** with the counterexample $5^2 - 1$.

73. The argument is a **valid** application of denying the conclusion.

75. **is a valid** example of rule A, affirming the hypothesis.

77. If two consecutive terms of a geometric sequence are 8 and 12, then r = 12/8 = 1.5 and each term can be calculated from the preceding term by multiplying by 1.5. So the sequence containing 8 and 12 would continue 18, 27, ... **Kevin is correct.**

79. Responses will vary among students. Some responses might include: If a person works for Sleezy, then that person will make a fortune; If you work hard, then you will get promoted fast; If you are trained by Sleezy, then you will be smart.

81. Responses will vary. They may include the ideas: The ad promotes the following conditionals: 'If you want to feel your best, then take one Vigorous Vitamin each day' and 'If a person takes Vigorous Vitamins, then that person cares about her health'. Frieda mistakenly has assumed the converse of the first ad statement and has assumed the inverse of the second ad statement. Although Frieda has misinterpreted the ad, she has most probably reached the conclusions hoped for by the ad maker.

83. Heather will weigh 137 pounds. $150 - (4 + 6 \times 1.5) = 137$.

85. Responses will vary among students. An argument might be something like: The bugs are assuming the conditional that if a person or animal slides down the slide, then that person or animal will become ensnared in the web. Some small animal does slide down. Thus the animal is ensnared. The bugs did pull it off. The final conclusion is that the bugs will eat like kings.

87. First, use the complete diagonal to get the magic number of 34. Now, column 3 is missing one number which must make the sum of that column 34. The number is 11. Continuing in the same manner we find that row 2 is missing a 5, column 1 must have a 9, the second diagonal needs a 6. Now put a 12 in the column 4 blank and 15 in the row 4 blank. Complete the perfect square by placing a 3 in row 1. Now all rows, columns, and diagonals add to 34.

16	**3**	2	13
5	10	**11**	8
9	6	7	**12**
4	**15**	14	1

SECTION 1.3

1. Responses will vary among students. Can you do this mentally? With paper/pencil? Need a calculator?

3. a. The diagram shows that the numbers of boxes in consecutive rows form an arithmetic sequence beginning with 1 and with a common difference of 1. One could extend the sequence, adding all the terms, until the sum is 21.
   $1 + 2 + 3 = 6$. $1 + 2 + 3 + 4 = 10$. $1 + 2 + 3 + 4 + 5 = 15$. $1 + 2 + 3 + 4 + 5 + 6 = 21$.
   Thus there are **6** boxes in the bottom row.

   b. The diagram shows that when the sixth row is drawn, the number of boxes totals to 21 and the last row contains 6 boxes.

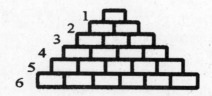

5. This problem has **one solution: 2 quarters, 2 dimes, and 1 nickel.** One might argue: since there must be 5 coins, none larger than a quarter, there must be at least 2 quarters because no combination of 4 or 5 nickels and dimes total to 75 cents. There cannot be 3 quarters because the total would be attained in 3 coins. So there must be 2 quarters. The remaining 3 coins must total 25 cents and this can be done in only one way: 2 dimes and 1 nickel.

7. The responses will vary among students. Possibilities include: make a table.

9. The responses will vary among students. Possibilities include: make a table.

11. The responses will vary among students. Possibilities include: use reasoning to match and eliminate possible matches.

13. Answers will vary among students. Possibilities include:
    a. Estimate to the next largest number of units of paint, probably gallons.
    b. Estimate to the nearest second.
    c. Estimate to the nearest $100 dollars.
    In none of these situations do you need an exact answer. You are not asked to reimburse a painter for materials, to determine the winner of the race, nor to pay for the vacation.

15. The responses will vary among students. The problem could be solved by guess-check-revise. Since the room is rectangular and has a perimeter of 44 ft, the sum of the length and width is 22 ft. If the room were 1 ft by 21 ft, the area would be 21 sq ft. If the room were 2 ft by 20 ft, the area would be 40 sq ft. If 3 ft by 19 ft, 57 sq ft. If 4 ft by 18 ft, 72 sq ft. Continuing in this fashion we find an area of 120 sq ft for a room **10 ft by 12 ft.**

17. One could reason that there are 4 choices for the chairperson and, for each one of these choices, there are 3 choices for secretary. So each of the 4 potential chairs could be matched with 3 potential secretaries. Thus there are 4 times 3, a total of **12** different officer combinations.

19. Responses will vary. One might reason the following way: At 4 AM the small hand is on the 4, the large hand on the 12. The minute hand will pass the hour hand as the clock goes to 5 AM, 6 AM, 7 AM, 8 AM, 9 AM, 10 AM, and 11 AM. The hands will be colinear at 12 noon, the passing point on the way to 1 PM. They will pass 1 final time on the way to 2 PM. Counting the passes, we see that the minute hand passes the hour and **9 times.**

21. Applying the guess-check-revise strategy we find that row 1 = **1 2 3** and row 2 = **5 6 4** is a solution to the problem.

23. Responses will vary among students. Possibilities include: Will it be cheaper for Nathan to install the TV himself or to have the store install it? What is the cheapest TV Nathan can buy and have the cost of the TV, the bracket, and the installation be cheaper than the cost of the TV and bracket and installing it himself?

25. #7   Suppose J is Jack's age. Then Chris is $J + 4$ years old. Together their ages total 30, so $J + (J + 4) = 30$ or $2J + 4 = 30$ or $2J = 26$, which means $J = 13$. So Jack is 13 years old, and Chris is $13 + 4$ or 17 years old.

    #8   As used in the solution of this problem, last refers to the terminal runner in a race. 'Second to last' is the same as 'next to last'. Now, one might draw a diagram and see that since Bella is fifth from last in a 21 person race, she is in 17th place. Now, she must pass one person to be in 16th place, a second person to be in 15th place, and so on until she has passed 14 persons and is in third place.

    #9   Write a few terms of the club membership and look for a sequence. The numbers joining are 1, 2, 3, 4, ..., 40. These numbers are in an arithmetic sequence. We notice that the 1st and 40th terms add to 41 as do the 2nd and 39th, as do the 3rd and 38th, and so on to the 20th and 21st. Thus there are 20 sums of 41, a total of **820 members.**

    #10   There are 4 choices for the first letter in the in the arrangement and for each of these there are 3 choices for the second letter. For each of these 12 choices for the first two letters there are 2 choices for the third letter. Thus there are a total of **24 possible arrangements.**

    #11   One might reason as follows: since Beth coaches a water sport and there is only one water sport, swimming, **Beth coaches swimming**. Because we know Anton's sister coaches soccer and because Diedre is the only remaining woman, **Diedre** must be Anton's sister as well as **the soccer coach.** We know **Anton** does not coach basketball so he must coach the remaining sport, **volleyball,** leaving **basketball to Cal.**

    #12   Pam's initial balance was $300. She then had a charge of $5 and wrote checks for $20, $50, $75, and $15 or a total of $160. So her balance was $300 − $5 − $160 = $135. After a deposit her balance was $200, so $135 + _____ = $200. She deposited $65.

27. Responses will vary among students. One might suggest to Angela that she is on the right track. She is getting closer to the answer, 24 to 9 is closer to 2 to 1 than is 21 to 6. She might increase her increment to 2 or 4 or 5 rather than the brute force of 1.

29. Pick any six balls and put three on each side of the balance. If this setup is in balance, use the balance to determine which of the two remaining balls is the heavier. If the original setup is not in balance take any two of the three balls on the heavier side and put them on the balance. If either is the heavier, it will move its side down. If the balance is level here, then the one of the 3 not on the balance scale is the heavier.

31. Responses will vary among students. A possible solution is: To fence a rectangular plot one needs posts at the corners. Suppose the posts are placed four meters apart. So: how many posts are needed on each long side? How many total on the long sides? How many posts are needed on each short side? How many posts total on the short sides? How many posts total so far? Are any posts counted twice? How many? How can you adjust the total for these double countings? (6, 12, 5, 10, 22, Yes, 4 corner posts, subtract 4, answer: 18)

33. a.   Consider any one of the five people. She will shake hands with 4 other persons. The total so far is 4 handshakes. Now consider a second person. There are 3 people remaining for her to shake hands with. The total is now 7 handshakes. Similar arguments give 2 handshakes for person 3, 1 handshake for person 4, and person 5 has already shaken hands with the other four guests. So the

total number of handshakes is $4 + 3 + 2 + 1 = \mathbf{10}$. The solution to the second problem has the same logic. Simply replace persons with dots and handshakes with line segments.

b. Responses will vary among students.

CHAPTER 1 REVIEW EXERCISES

1. Answers will vary among students.

3. Responses will vary. Possibilities include:
    a. A counterexample is $10 \div 1 = 10$.
    b. A rectangle may have sides that are not equal, say 3 and 4. Thus it is not a square.
    c. Consider the product of 7 and 2. The product is 14, an even number.

5. The conclusion, obtained by **affirming the hypothesis,** is that angle C of figure ABCD is a right angle.

7. Responses will vary among students. Some examples are:
    a. My aunt lives in Austin. Therefore she lives in Texas.
    b. I don't know anyone who lives in Texas. Thus I know no Austinians.
    c. Since Joe lives in Texas, Joe lives in Austin.
    d. Since I do not live in Austin, I don't live in Texas.

9. The conjunction p and $\sim p$ is false because a conjunction requires that both parts be true. However, p and $\sim p$ cannot be true simultaneously.

11. Converse: If you live in LA, you live in California. (True)
    Inverse: If you do not live in California, you do not live in LA. (True)
    Contrapositive: If you do not live in LA, you do not live in California. (False)

13. a and b. i. 22, 33, 44, 55, 66, 77, 88, 99, **110, 121, 132: arithmetic, d = 11.**
    ii. 1, 11, 12, 22, 23, 33, 34, 44, **45, 55, 56: neither**, sequence is formed by alternately adding 10 and 1
    iii. 2, 6, 18, 54, 162, 486, 1458, **4374, 13,122, 39,366: geometric, r = 3.**

15. a. The first column is an arithmetic sequence with $d = 1$. The second column is an arithmetic sequence with $d = 2$. Continuing these sequences we get the next three rows as: **7 13; 8 15; and 9 17.**
    b. We also note that the entries in the second column are 1 less than twice the corresponding first column entries. So the 25th row would be **25 49.**

17. Responses will vary. A possibility is:
    a. Use a guess-check-revise strategy.

b.   Parking–$10, concert ticket–$40, total $50 is low. So revise the cost of the parking up. Parking–$15, concert ticket–$60, total $75 is still low. So revise the cost of the parking up. Parking–$18, concert ticket–$72, total is $90. So parking cost **$18, and the concert ticket cost $72**.

19. These two problems are essentially the same because they can be represented by a common geometric model. The vertexes in the model can represent either the people or the desks, the segments can represent the cables or a pairing of riders.

21. The points A, B, and C represent the cities. The arcs and diameters represent the roads.

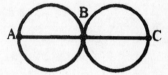

23. A possibility is the circle graph:

25. a.   The common aspect of the figures is that the circle must be inside the square. The orientations of the square with respect to the circle may differ.

b.   Responses will vary. A set of possibilities is:

i   ii   iii

27. Fill in the following table:

Number of Items	Cost Up Front Plan	Straight $40 Cost
0	100	0
1	120	40
2	140	80
3	160	120
4	180	160
5	200	200

The breakeven point is **5 items purchased.**

29. Answers will vary among groups. Problem-solving skills refer to the ability of individuals or groups of persons to implement the problem-solving process with the end result of arriving at suitable strategies to solve particular problems.

31. Caitlin is correct. She could pass the same person more than once and someone she passed could repass her to finish ahead of her.

SECTION 2.1

1. A one-to-one correspondence can be established between two sets if they have the same number of elements. Sets **B and E** both have 4 elements. Both sets **A and D** have zero elements. They are different representations of the null set.

3. $n(S)$ represents the number of elements in a set.
    a. There is **1** element in the set {yellow].
    b. There are **0** elements in the set $\varnothing$, the null set.
    c. There are an infinite number of elements in the set {3, 6, 9, 12, ...}.
    d. There are **5** elements in the set {1, 2, 3, 4, 5}.
    e. There is **1** element in the set {$\varnothing$}.

5. A set S1 is a subset of a set S2 whenever every element of S1 is also an element of S2. A set S1 is a proper subset of a set S2 whenever every element of S1 is also an element of S2 and S2 also contains elements not in S1. Every set is a subset of itself, but not a proper subset. The null set is a proper subset of all sets except itself. It is a subset of itself. Applying these concepts we see that:
    a. **A** is not a proper subset and not a subset **C; B** is not a proper subset and not a subset **A; B** is not a proper subset and not a subset **C**.
    b. No sets meet this conditions. A proper subset is a subset.
    c. A, B, and C are all subsets of themselves. So pairs **A-A, B-B, C-C** meet the conditions.
    d. Any proper subset is also a subset. So **A-B, C-A, C-B** meet the conditions. A, B, C, are all proper subsets of the universal set U.

7. Answers will vary. $S \subseteq T$ means that S is either a proper subset of T or else equal to T. $S \sim T$ means that S and T have the same number of elements. Together these imply that $S = T$. $R \subset S$ means that R is a proper subset of S: all elements of R are in S but not all elements of S are in R. So a solution is: $R = \{a,b\}$, $S = \{a,b,c\}$, $T = \{a,b,c\}$.

9. a. The set $Z = \{z\}$ is a subset of the set $S = \{w,x,y,z\}$ because all of the elements of Z are contained in S.
    b. The set $A = \{a\}$ is not a subset of S because there is at least one element of A that is not in S.

11. a. $S = \{w,x,y,z\}$ is a subset of itself because there is no element of S that is not an element of S.
    b. The argument may be applied to any set S.

13. a. Cardinal–tells how many
    b. Nominal–tells which one; Ordinal–tells what place
    c. Nominal–tells which one
    d. Ordinal–tells what place

15. a. $S = \{0, 1, 2, 3, 4, 5\}$. This is a finite subset of the whole numbers and it does have a greatest element, 5.
    b. $T = \{6, 7, 8, 9, ...\}$. This subset of the whole numbers does not have a greatest element.
    c. There is no infinite subset of the whole numbers that has a greatest element.

17. The subsets of {red, green, blue} are:{ }, **{red}, {green}, {blue}, {red, green}, {red, blue}, {green, blue}, {red, green, blue}.**

19. All sets that are in a one-to-one correspondence are defined as equivalent sets and are associated with the same whole number.

21. The sets A = {a,b,c,d}, B = {w,x,y,z}, and C = {a,b,y,z} are all equivalent to the set S = {a,b,c,d}.
    a. Sets C and S show that equivalent sets do not necessarily have all elements in common.
    b. Sets B and S show that equivalent sets do not necessarily have any elements in common.
    c. Equivalent sets must have the same number of elements.

23. a. The empty set is the only subset of itself.
    b. The empty set has no proper subsets because it is the only subset of itself;
    c. One can conclude that there is no whole number less than the number associated with the empty set, the number 0.

25. a. **False**: equivalent sets are not necessarily equal. Consider A = {a,b} and B = {c,d}.
    b. **True**: if two sets do not contain the same number of elements they cannot contain exactly the same elements.
    c. **True**: if two sets contain exactly the same elements those sets contain the same numbers of elements
    d. **False**. Consider A = {a,b} and B = {c,d}.

27. The librarian means that faculty may check out only some, not all, of the newly acquired books.

29. The elements of the sets W and N may be paired: 0-1, 1-2, 2-3, 3-4, 4-5, ..., $(n-1)$-n, n-$(n+1)$, ...

31. Let the persons available to serve on the committee comprise the set S = {A, B, C, D, E}. A subcommittee must contain at least two persons and may be a committee of the whole. So the number of possible subcommittees is the number of subsets of S excluding the empty set and the 5 singles: $2^5 - 1 - 5 = 32 - 6 = 26$.

33. Responses will vary. Possibilities include: sets of students and sets of desks, persons invited to a party and party favors, plates and persons at the dinner table. Some problems are trivial such as more Christmas cards than stamps; others more serious as more children than doses of vaccine.

SECTION 2.2

1. a. The union is E = {those people who are either more than 20 years old or those persons enrolled in college (or both)}.
    b. The union is C = {2, 4, 6, 8}
    c. The union is G = {1, 3, 5, 7, 9, ... , 37, 39, 41, 42, 43, 44, ...}.

3. a. The single element common to both sets is e. So R ∩ S = **{s, t, p}.**
    b. The set consisting of the elements in both R and T is **{s, t, o, p, l, i, e, n}**.
    c. The set consisting of the elements in both S and T is **{p, l, i, s, t, e, n}**.
    d. (R ∩ S) ∩ T = {s, t}.
    e. R ∩ (S ∩ T) = {s, t}

5. a.   **0   4   9**

    b.   **0   6**

    c.   **0   10 12**

7. If 2000 miles were traveled on the interstate and 356 miles on secondary roads, what total distance was covered?

9. Examples will vary.
    a. This is an application of the **commutative** property of addition: $2 + 8 = 8 + 2$.
    b. This is an application of the **identity** property of addition: $a + 0 = 0 + a = a$.
    c. This is an application of the **associative** property of addition:
        $8 + 7 = 1 + 7 + 7 = 1 + (7 + 7) = 1 + 14 = 15$.
    d. This is an application of the **uniqueness** property of sums.

11. a. {♦, ♦, ♦, ♦, ♦, ♦, ♦, ♦, ♦}
    b. {♣, ♣, ♣, ♣, ♣, ♣}
    c. {1, 2, 3̸, 4̸, 5̸, 6̸, 7̸, 8̸, 9̸, 1̸0̸, 1̸1̸, 1̸2̸}.

13. a. $3 + 1 = 4$ $4 - 1 = 3$
        $1 + 3 = 4$ $4 - 3 = 1$
    b. $15 + 10 = 25$ $25 - 10 = 15$
        $10 + 15 = 25$ $25 - 15 = 10$

15. The definition of subtraction is: $a - b = c$ if and only if there exists a unique whole number, c, such that $a = b + c$. (a) $18 = n + 8$ (b) $25 = 15 + x$ (c) $y = 129 + 83$ (d) $a^2 = 30 + a$

17. A student had 12 lottery tickets. Five of these were known losers. How many tickets remained?

19. A professor gives a quiz with 12 questions. If 5 of the questions are essay questions, how many of the questions are not essay questions?

21. The professor has written 5 essay questions for a quiz. How many more questions does the professor need to write so that the quiz will have 12 questions?

23. a.   The key sequence would be: $+ \mathbf{1} = = + \mathbf{10} = = = = + \mathbf{100} = = =$.
    c.   The display would be: 1, 2, 12, 22, 32, 42, 142, 242, 342.
    d.   The properties of **commutativity and associativity** are applied.

25. a.   The set of odd whole numbers is not closed under addition. For example, $3 + 5 = 8$.
    b.   The natural numbers are a subset of the whole numbers that lacks an additive identity element.
    c, d. These subsets of the whole numbers do not exist.

27. a.   The properties used are associativity and base-ten place value: $9 + 6 = 9 + (1 + 5)$
         $(9 + 1) + 5 = 10 + 5$. But in base-ten, 10 is represented by a '1' in the second position. Thus
         $10 + 5$ is represented as 15.
    b.   The response to this question depends upon the insights of the students. A possibility is: the sums
         of $1000 + 500$, $100 + 50$, and $10 + 5$ have the same non-zero digits.

29. $a < b$ if and only if a is to the left of b on the number line.

31. Responses will vary among students. Students might point out that subtraction problems can be rewritten as missing addend addition relations. Separating a set into two sets is modeled by physically removing objects from some set. The objects removed may be placed into a one-to-one correspondence with some third set. Thus 2 sets are formed: one from the objects removed and the other by the objects that remain. When comparing sets one may remove objects from the two sets in pairs, one from each set, to determine which, if either, has elements remaining after one of the sets has been reduced to the null set.

33. a.   If a increases or c decreases, then e increases. When a decreases or c increases, then e decreases. If a increases and c decreases, e increases.
    b.   Tests applied will vary among students.
    c.   See part a.
    d. e decreases if: a decreases, c increases, a increases by less than c increases, a decreases by more than c decreases. e increases if: a increases, c decreases, a increases by more than c increases, a decreases by less than c decreases.
    e.   Responses depend on interactions among students.

35. **False.** The set of odd numbers is an infinite set not closed under addition because $1 + 3 = 4$.

37. Responses may vary. One possibility is:

    This system is analogous to addition mod 4.

#	a	b	c	d
a	a	b	c	d
b	b	c	d	a
c	c	d	a	b
d	d	a	b	c

39. a. If the condition $k > 0$ were omitted but we still stay in the set of whole numbers, we have the situation that although a equals a, $a + 0 = a$ which implies that a is greater than a.
    b. The definition of 'less than' might be stated: a is less than b if and only if there exists some whole number $k > 0$ such that $a = b - k$.

41. Responses will vary. Possible ideas are: Scribes might have treated subtraction as missing addend problems. Suppose the scribe wanted to determine the difference $14 - 6$. Interpreting the problem as $14 = 6 + ?$ the scribe would locate that 14 in the body of the table that had one addend of 6 and find the corresponding other addend.

SECTION 2.3

1. To verify the operations one would:
    a. show the union of 8 disjoint sets each with four elements.
    b. show the union of 5 disjoint sets each with 10 elements.
    c. show the union of 6 empty sets.

3. a.
    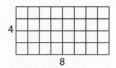

    b.

    c. The is no array of one dimension equal to 0.

5. If 4 students are running a relay in which each student runs 30 meters, how long is the race?

7. a. 4 chairs • 6 fabrics = **24** different chairs

9. w wrapping paper types $\times$ c bow colors = 64 paper-bow combinations

11. a. Multiplication is commutative: $a \times b = b \times a$.
    b. There is a **zero property** of multiplication: for each whole number a, $a \times 0 = 0 \times a = 0$.
    c. In the whole numbers multiplication is distributive over addition. Suppose that $c + 1 = b$. Then $a \times b = a \times (c + 1) = ac + a$.
    d. The whole numbers are **closed** under multiplication: the product of 2 whole numbers is a unique whole number.
    e. Associative property of multiplication

13. A teacher has a bank of 24 test questions and wishes to create tests with 6 questions. How many tests are possible?

15. A teacher has a bank of 24 test questions and wishes make 6 tests with the same number of questions. How many questions will be on each test?

17. A teacher made 6 tests with the same number of questions. The total number of questions was 24. How many questions were on each test?

19. a. $18 = 6 \times n$; $n = \mathbf{3}$.
    b. $25 = 5 \times X$; $X = \mathbf{5}$.
    c. $y = 42 \times 126$; $y = \mathbf{5292}$.
    d. $0 = b \times c$. $\mathbf{c = 0}$ (b cannot be 0 because it is the denominator of the original expression. Thus b is any natural number.)

21. a. $26 \div 3 = \mathbf{8\ r\ 2}$ because $8(3) + 2 = 26$.
    b. $292 \div 21 = \mathbf{13\ r\ 19}$ because $21(13) + 19 = 292$.
    c. $4 \div 7 = \mathbf{0\ r\ 4}$ because $0(7) + 4 = 4$.

23. Predictions will vary among students.

predicted n(C)	C	actual n(C)
a.	{ } (there are no elements in { } to pair with elements of the second set. Thus there are no ordered pairs and C is empty.)	**0**
b.	{(r, a), (r, b), (s, a), (s, b), (t, a), (t, b)}	**6**
c.	{(a, 5), (a, 6), (a, 7), (a, 8), (a, 9)}	**5**

    d. $4 \times 0 = 0$; $3 \times 2 = 6$; $1 \times 5 = 5$.

25. To determine the value of a particular place within the numeral the appropriate power of 10 is multiplied by the face value of that place.

27. a. $\mathbf{4r(5 + 4)}$.
    b. $\mathbf{9(3a^2 + 9 + 1)}$.
    c. $\mathbf{2(12a + 7b + 10c)}$.

29. Responses will vary among students. They might be based upon the concept that division problems can be rewritten as 'missing factor' problems. The table can be used to find the missing factor.

31. a.   If the values of q and r that satisfy $(q \times 12) + r = 64$ are restricted to the whole numbers, the (q, r) pairs are: (0, 64), (1, 52), (2, 40), (3, 28), (4, 16), (5, 4).
    b.   Only the pair $q = 5$, $r = 4$ satisfies the division algorithm because when dividing 64 by 12 the remainder must be less than the divisor.
    c.   If the restriction were dropped any of the above (q, r) pairs would satisfy the division algorithm and the quotient would not be unique.

33. The responses to this questions will vary among students and will depend on the insights developed by the students. Students may notice that the table is symmetric with respect to the upper-left to lower-right diagonal because of the commutative property.

35. **Yes, it could be true.** When there were 4 candy dishes the possible remainders were 0, 1, 2, 3. If these occurred with equal frequency, then Eleanor ate an average of one and a half pieces of candy for each bag. But with 7 dishes, the possible remainders are 0, 1, 2, 3, 4, 5, 6 which gives an average of 3 candies left after filling the dishes.

37. **All these sets except (a) are closed under multiplication**. First consider the set of even whole numbers. 0 times any element is 0 and therefore is in the set. A whole number is even if and only if it has a factor of 2. Thus the product of 2 even numbers is even because the product has a factor of 2. A similar argument holds for a set of multiples of 10. Now consider the odd whole numbers. An odd number is an even plus 1: $O = E + 1$. So the product of 2 odd numbers is: $(E + 1) \times (E + 1) = EE + (E + E) + 1$. Now, EE is even and the sum of evens is even. So $EE + E + E$ is even. Adding 1 to this sum results in an odd number. For parts b and d we see that $1 \times 1 = 1$ and $0 \times 0 = 0$.

39. a. $0 \div a = 0$ because this is equivalent to $0 = a \times 0$ which is true for all a not equal to 0 by the definition of the zero property.

   b.   Suppose $a \div 0$ were defined and had a quotient c. We could rewrite it as a multiplication relation: $a \div 0 = c$ becomes $a = c \times 0 = 0$. But c could be any whole number and a quotient, by definition, is unique.

   c.   Suppose $0 \div 0$ were defined and had a whole number quotient c. Then, by definition of division and quotient, c is a unique whole number. So $0 \div 0 = c$; or $0 = c \times 0$. But this would be true for all whole numbers. But the quotient of two whole numbers is unique. Thus $0 \div 0$ is undefined.

41. This question requires student interaction and activity.

43. To bring closure to the whole numbers under division we must expand the set of numbers to include the rational numbers. They are useful in situations in which a unit is to be distributed to a number of individuals. The whole numbers are a subset of the rational numbers. The quotient of any 2 whole numbers, if defined, exists within the set of rational numbers.

SECTION 2.4

1. a.   **numeral**
   b.   **number**
   c.   **numerals**

3. a.   The number 586 would be represented with 5 'hundreds' squares, 8 'tens' sticks, and 6 'units' cubes.
   b. $4392 = \mathbf{4(1000) + 3(100) + 9(10) + 2(1)}$.
   c. $2,864,071 = \mathbf{2(10^6) + 8(10^5) + 6(10^4) + 4(10^3) + 0(10^2) + 7(10^1) + 1(10^0)}$.

5. a. $3(36) + 4(6) + 4(1) = \mathbf{136}$.
   b. $2(16) + 0(4) + 2(1) = \mathbf{34}$.
   c. $1(32) + 1(16) + 0(8) + 0(4) + 1(2) + 1(1) = \mathbf{51}$.

7. a. $42_{seven} = 4(7) + 2 = 30_{ten} = 1(25) + 1(5) + 0 = \mathbf{110_{five}}$.
   b. $32_{four} = 3(4) + 2 = 14_{ten} = 2(6) + 2 = \mathbf{22_{six}}$.
   c. $1101100_{two} = 1(2^6) + 1(2^5) + 1(2^3) + 1(2^2) = \mathbf{108_{ten}} = 1(4^3) + 2(4^2) + 3(4^1) + 0(4^0) = \mathbf{1230_{four}} = 1(8^2) + 5(8^1) + 4(8^0) = \mathbf{154_{eight}}$.

9. As described in the text: "...a tally system is based on establishing a one-to-one correspondence between a single mark and a single object so that the marks represent the number of objects. Later, grouping was used to simplify numeration systems."

11. a. **1272**;   b. **35**;   c. **11,461**;

13. a. **126**;   b. **92**;   c. **510**.

15. Various students will approach this question with different degrees of rigor.
    a. Egyptian numeration is based on a grouping system of powers of ten. One of each power of ten is represented by a different symbol and a number is represented with the fewest number of symbols. The Hindu-Arabic system is also based on powers of ten but is a place value rather than a grouping system.
    b. All students should note that both the Babylonian and Hindu-Arabic systems are place value systems although they use different bases. The Hindu-Arabic system has a different symbol for each numeral representing numbers from zero through ten. The Babylonian numeration system uses a modified tally and grouping system within the places.
    c. There are few similarities between the Roman and the Hindu-Arabic numeration systems. The Roman system is unique in that it utilizes both additive and subtractive relations. The Roman system also uses a grouping symbolism similar to the Egyptian.
    d. Hindu–Arabic has 10 symbols, but Mayan has only 3; Mayan uses tallying, but Hindu–Arabic doesn't.

17.

BASE-TEN NUMERALS

1	2	3	4	5	6	7	8	9	10
11	12	13	14	15	16	17	18	19	20
21	22	23	24	25	26	27	28	29	30
31	32	33	34	35	36	37	38	39	40
41	42	43	44	45	46	47	48	49	50
51	52	53	54	55	56	57	58	59	60
61	62	63	64	65	66	67	68	69	70
71	72	73	74	75	76	77	78	79	80
81	82	83	84	85	86	87	88	89	90
91	92	93	94	95	96	97	98	99	100

    a. To find 10 more than a number, go down one row in the same column.
    b. To find 10 less than a number, go up one row in the same column.
    c. To find one more than a number, move one space to the right or, if at the end of a row to the first space in the next row.
    d. To find one less than a number, move one space to the left or if at the left end of the row, go to the far right space in the above row.
    e. To find 11 more than a number, move one row down and one space right.
    f. To find eleven less than a number, move one row up and one space left.
    g. To find 9 more than a number, move one row down and then one space left.
    h. To find 9 less than a number, move one row up and then one space right.

19. The numeral $rstu_b$ may be expanded to: $r(b^3) + s(b^2) + t(b^1) + u(b^0)$.

21. a. The count of each set of symbols was multiplied by the value of the symbol. These products were then summed to obtain the base-ten numeral.
    b. The numerals in each place were summed as in part a and then these values were multiplied by the appropriate power of 60. Finally, these values were summed to obtain the base-ten value.
    c. The Roman numerals were grouped into sets of the same symbol and pairs of different symbols. If the different symbols decreased left to right their values were added. If the different symbols increased left to right the smaller was subtracted from the larger. Values of groups of the same symbol were multiplied by the number of symbols in the set. Finally, all these values were summed to obtain the base-ten representation.
    d. Determine the place of each symbol, multiply to determine the value of each symbol, and then add all the values.

23. It could be a square of 1000×1000 cm, or it could be a cube of $100 \times 100 \times 100$ cm.

25.   BASE-FIVE CHART   BASE-TEN CHART
      1   2   3   ·   10
     11   12   13   14   20   SEE
     21   22   23   24   30   THE BASE-TEN CHART
     31   32   33   34   40   IN PROBLEM 13
     41   42   43   44   100
The patterns identified in the two charts should be essentially the same.

27. The responses to this question depend upon the imaginations of the students.

29. Assume that there are 10 hamburgers to a foot. Then 400 billion hamburgers would form a stack about 40 billion feet high. Assuming 5000 feet per a mile, the 40 billion (40 000 000 000) feet high stack is about 8 million miles high. This distance is about 32 times the earth-moon distance or about one twelfth the earth-sun distance. So the correct answer is **e**.

31. Responses to this question depend on the prior mathematical and educational experiences of students and upon their willingness to do research.

33. The base-ten number represented by the numeral 421 could be represented as **645$_{eight}$**; that is 6 groups of 64 plus 4 groups of 8 plus 5 units. If the base is to be 16, then additional symbols are required. A base 16 system similar to base ten requires 16 separate symbols. These commonly are, in counting order, 0, 1, 2, 3, 4, 5, 6, 7, 8, 9, A, B, C, D, E, F. Thus 421 = **1A5$_{sixteen}$**. This numeral is I group of 256 plus A = 10_{ten} groups of 16 plus 5 units.

CHAPTER 2 REVIEW EXERCISES

1. a. The following sets have '7' in common: {#,#,#,#,#,#,#}. {<,<,<,<,<,<,<}, {*,*,*,*,*,*,*},
        {$,$,$,$,$,$,$}, {1, 2, 3, 4, 5, 6, 7}.
    b. The sets { }, { }, { } show the meaning of 0.
    c. Let N = {$,$,$,$,$,$,$,$,$} and E = {@,@,@,@,@,@,@,@,@,@,@,@,@,@,@}. Since N is
        equivalent to a proper subset of E, $n(N) < n(E)$ or $9 < 15$, $15 > 9$.
    d. Since { } is equivalent to a proper subset of any non-empty set, $0 <$ all other whole numbers.

3. Responses will vary among students. Responses might include:
    a. The Egyptian numeration system is a tally system through 9 of a symbol with a different symbol
        for every power of 10. Early numeration had no 0 symbol.
    b. The Babylonian system is a place value system based on powers of 60. It incorporates a tally
        system of 2 symbols tallying units to 9 and tens to fifty.
    c. The Roman system uses different symbols for units, tens, fifties, hundreds, five hundreds, and
        thousands. It employs both additive and subtractive characteristics to keep the number of
        repetitions of a symbol to 3 or less.
    d. The Mayan System is based on groups of 20, some tallying, place value, and 3 symbols,
        including a zero.
    e. The Hindu-Arabic system is a place value system based on powers of 10. It uses different
        symbols for the numbers 0 through 9.

5. a, b. Student activities. For example, to represent 2045 use 2 thousands cubes, no flats, 4 sticks and 5
        units.
    c. $2045 = 2(1000) + 0(100) + 4(10) + 5(1)$.
    d. $2045 = 2(10^3) + 0(10^2) + 4(10^1) + 5(10^0)$.

7. Responses will vary. Examples are:
    a. I have 85 cm of heavy-gauge wire and 62 cm of fine-gauge wire. How much wire do I have?
    b. I have 85 National League baseball cards and 62 American League baseball cards. How many
        baseball cards do I have?

9. Responses will vary. Examples are:
    a. There are 12 rows in my garden and each row is 25 feet long. How many feet of soaker hose do I
        need to water the entire garden at once?
    b. Each of 12 students brings 25 candies to class. How many candies were brought to class?
    c. A rectangle is 12 by 25 inches. What is the area of this rectangle?
    d. Joe has 12 shirts and 25 ties. How many different combinations of shirt and tie may he select
        from?

11. The set of flowers may be represented {a,b,c,d}. A student may use some or all of the flowers, a
    subset of the set of flowers, but no student may use no flowers in an arrangement. There are 15
    subsets of a 4 element set, excluding the null set. So the 15 students can each make a different
    arrangement.

13. $180 - 59 = 121$ because $180 = 59 + 121$.

15. a. **S is closed** under addition because the sum of multiples of 10 is a multiple of 10.
    b. The **additive identity is 0** because any element of the set added to 0 results in the original element.
    c. The set is **closed under multiplication** because the set is multiples of 10 and if a multiple of 10 is multiplied by any whole number not 0, then the product contains a factor often and thus is a multiple of 10.
    d. The set **does not have** an element such that multiplication of any element by this elements results in the original element.
    e. The **associative, commutative, and distributive properties do apply** to S.

17. $125 \div 40 = 3$ r 5 because $3 \times 40 + 5 = 125$.

19. The drawings will vary among students. They should represent the relation that sets are in a one-to-one correspondence if the elements of one set can be paired with elements of another set. If the pairing can produce pairs of identical elements the sets are called equal.

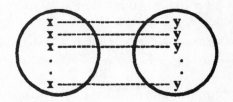

          X   **Y**

If each element of X can be paired be paired with an element of Y then the sets are in a 1 to 1 correspondence, represent the same whole number, and are equivalent. If, further, each element of X can be paired with an identical element in Y, the sets are equal.

21. Responses this question depend upon interactions within groups of students.

SECTION 3.1

1. Compatible pairs of numbers are 12 and 18 (sum = 30 by inspection), 46 and 64 (sum = 110 by inspection). So, applying the commutative and associative properties of addition one gets:
    $12 + 46 + 18 + 64 = 12 + 18 + 46 + 64 = 30 + 110 = \textbf{140}$.

3. As in part #2, one gets: $25 \times 28 \times 4 = (25 \times 4) \times 28 = 100 \times 28 = \textbf{2800}$.

5. $2579 - 372 = (2500 - 300) + (79 - 72) = 2200 + 7 = \textbf{2207}$.

7. $48 \cdot 20 = 48 \cdot 2 \cdot 10 = (48 \cdot 2) \cdot 10 = 96 \cdot 10 = \textbf{960}$.

9. To compute $848 - 300$ one need only count back by one hundred three times: 848, 748 (back 1), 648 (back 2), 548 (back 3). So the result is **548**.

11. To compute $648 + 32$, first one counts on by ten three times and then twice in the units place: 648, 658, 668, 678, 679, **680**.

13. To calculate $75 - 39$, first calculate $75 - 40$, then add 1: $75 - 39 = (75 - 40) + 1 = 35 + 1 = \textbf{36.}$

15. To calculate $19 \cdot 8$: $19 \cdot 8 = 20 \cdot 8 - 8 = 160 - 8 = \textbf{152}$.

17. Apply the associative property: $(85 + 12 + 18) = 85 + (12 + 18) = 85 + 30$. Now count by ten three times: 85, 95, 105, **115**.

19. Reverse the distributive property [factor]: $(18 \cdot 6) - (15 \cdot 6) = (18 - 15) \cdot 6 = 3 \cdot 6 = \textbf{18.}$

21. Techniques for the computation will vary. One possible method: the top four category values are: 930, 900, 850, 700. The lowest four are 100, 130, 135, 193. We want to perform the computation: $D = (930 + 900 + 850 + 700) - (100 + 130 + 135 + 193)$. We can apply the commutative and associative properties to rewrite the expression for D: $D = 930 + 900 + 850 + 700 - 100 - 130 - 135 - 193$ $= (930 - 130) + (900 - 100) + (850 - 135) - 193 + 700 = 800 + 800 + 715 - 193 + 700$. Then using the equal addition technique on the 3rd and 4th terms we have: $800 + 800 + 722 - 200 + 700$. Counting on and back in the hundreds place gives $1600 + 722 + 500 = 2100 + 722 = \textbf{2822}$.

23–32. The techniques will vary among students. Some possible techniques are:

23. Associative property and compatible numbers: $12 \cdot (40 \cdot 5) = 12 \cdot 200 = \textbf{2400}$.

25. Break apart, commutative and associative properties:
    $53 + 26 = 50 + 3 + 20 + 6 = (50 + 20) + (3 + 6) = 70 + 9 = \textbf{79}$.

27. Commutative and associative properties and compatible numbers:
    $24 + 39 + 76 = (24 + 76) + 39 = 100 + 39 = \textbf{139.}$

29. Use compensation: add 1 then subtract 1. $147 - 38 = 147 + 1 - 38 - 1 = 148 - 38 - 1 = 110 - 1 = \mathbf{109.}$

31. Use compensation: add 1 then subtract. $455 - 26 = 455 + 1 - 26 - 1 = 456 - 26 - 1 = 430 - 1 = \mathbf{429}.$

33. The student is including the starting number (410 or 2100) as the first number when counting back. Instead of saying "410-400-390," he should say "400-390-380."

35. Responses will vary among students. A possible response is:
$7 + 9 + 3 + 1 = 7 + 3 + 9 + 1 = 10 + 10 = 20.$

37. Responses will vary among students. A possible response is: 25 is $20 + 5$. So $11 \cdot 25$ is $11 \cdot 20$, which is 220, plus $11 \cdot 5$, which is 55. So the product is $220 + 55 = \mathbf{275}.$

39. $45 + 63 + 25 + 17 = (45 + 63) + (25 + 17) = 45 + (63 + 25) + 17 = 45 + (25 + 63) + 17 = (45 + 25) + (63 + 17) = 70 + 80 = 150$

41. The increase is greater for private colleges. The increase for private colleges is $\$29,026 - \$6,026 = \$13,000.$ The increase for public colleges is $\$12,127 - \$6,825 = \$5,302.$

43. The total number of medals is $76 + 72 + 67 + 59 + 59 = 333.$ The mental math strategies students use will vary.

45. The total price for a queen set and a twin set from W. S. Manufacturing is $\$119 + 2(\$33)$ and the total price for the same two sets from Royalty Premier is $\$499 + 2(\$179)$. So: $(499 + 358) - (119 + 66) = (500 - 120) + (362 - 70) = 380 + (370 - 70) - 8 = 680 - 8 = 672.$ The difference in total price is $\mathbf{\$672}.$

47–51. Answers will vary among students. Possible responses are:

47. First use an under-estimate: 28 times 20 is 560, more than 500. So the answer is 'yes'.

49. An exact answer is called for. The data is such that mental computation by break-apart into compatible numbers is an appropriate method.
$34 + 46 + 52 + 38 = (30 + 40) + (50 + 30) + (4 + 6) + (2 + 8) = 170.$

51. Because the actual questions asks "… about how much…", only an estimate is required. Mental computations are appropriate. Estimate the cost per tire as $50.

53. Responses will vary among students. A representative response is: For a fund-raiser dinner for funds for cancer research, 48 tables of six have been pledged. How many dinners must be provided? The problem can be solved: $(48)(6) = (50)6 - 2(6) = 300 - 12 = 300 - 10 - 2 = 290 - 2 = \mathbf{288}.$

55. The responses to this question will vary.

57. Responses will vary among students. Possible responses are:
.   A similar symbolism might permit the appending of a '5' symbol to each numeral producing an equivalent but simpler computation: $2940 - 1500.$

SECTION 3.2

1. 6783 rounds to **6780**. Since 3 is < 5, replace it with a 0.

3. 20**9**.8 rounds to **210**. Since 8 ≥ 5, the place to the left is incremented.

5. **555** rounds to **600,** since 5 ≥ 5.

7–10: Responses will vary among students. Possible estimates include:

7. Use rounding to replace 4671 + 2509 with 5000 + 2500 and obtain the estimate **7500**.

9. Replace 864 × 22 with 900 × 20 to obtain the estimate **18,000.**

11–13: Responses will vary. Possible responses are:

11. Replace 424 + 526 with 425 + 525 and obtain the result **950**.

13. Replace 23 · 8 with 20 · 10. The estimated result is **200**.

15. 433 + 126 + 678 + 400 can be approximated by 400 + 100 + 600 + 400 = **1500.**

17. 1286 ÷ 63 can be approximated by 1200 ÷ 60 to obtain the estimate **20**.

19–21: Answers may vary among students. Possible responses include:

19. 824 + 238 is approximated by (800 + 200) + (20 + 40) = 1000 + 60 = **1060.**

21. 38 + 64 + 46 + 76 + 87 can be approximated by (30 + 60 + 40 + 70 + 80) + (10 + 10 + 10) = **310.**

23–25: Responses will vary among students. Possible solutions are:

23. Since all the addends cluster around 100, we can approximate the sum as 5 × 100 = **500**. Since some are above 100 and others below 100, compensation is not necessary.

25. We can approximate the sum as 4 × 10 = **40** and compensate if desired.

27–31: Responses will vary. Student responses might include:

27. The product 54 × 38 is between 50 × 30 = **1500** and 60 × 40 = **2400.**

29. A range for the difference is (600 – 200) = **400** to (700 – 100) = **600.**

31. Responses will vary. One possibility is rounding: 325 × 42 is approximated by 300 × 40 = 12000. Since both factors were rounded down, this is an underestimate.

33–40: Responses will vary among students. Possible responses include:

33. 765 + 824 + 799: first round 799 to 800 then use front–end estimation with adjustment. So we have the sum approximated by (700 + 800 + 800) + (75 + 25) = **2400.**

35. Since front–end subtracts from both numbers and since the amount dropped from each of the numbers is approximately the same, front–end gives the approximation 8000 – 4800 = **3200.**

37. Use compatible numbers: $500 \div 100 = $ **5**.

39. Use rounding to thousands and the associative property:
24,000 + (19,000 + 11,000) + 8,000 = **62,000.**

41–45: Responses will vary among students. Possible responses include:

41. Using rounding: 25,000 + 66,000 = 91,000. The exact answer is greater. The estimate is low because one addend was rounded down more than the other was rounded up. The exact sum is **91,335**.

43. Using clustering we get an estimation of $4 \times 800 = 3200$. Because the numbers less than the cluster number are farther from the cluster number than those greater, the exact result, **3165**, is less.

45. Using rounding we estimate $700 \div 70 = 10$. But since 700 < 726 and 70 > 67, the exact answer, **10.8,** is greater than the estimate.

47. Responses will vary among students. Possible responses include: A possible addend is any number between 400 and 449. The estimator might have used the rounding technique. Another possibility for the addend is any number between 500 and 599. The estimator might have use the first digit front-end technique. Another possible addend is a number similar to 432. The estimator might have used 450 as a cluster number.

49. The estimator employed the techniques of approximating the computation by **clustering and then rounding.**

51. Harvey knows that 8,000 – 3,000 is 5,000 and notes that the difference between 8,000 and 7,653 is greater than the difference between 3,000 and 2,861. So the over-estimate of 7,653 is greater than the over-estimate of 2,861. Over-estimating the minuend produces a large estimate, over-estimating the subtrahend produces a small estimate. Since the over-estimate of the minuend is greater than the over-estimate of the subtrahend, the overall **estimate (5,000) is larger** than the exact difference.

53. Responses will vary among students. Possible responses include: the sum of 238 and 273 is approximately 500. Use 250 as a cluster number. The product of 21 and 26 is approximately 500. Use the compatible numbers 20 and 25.

55. Possible response: Rounding to the leading digit in each calculation gives $700 \div 90$ and $3,000 \div 40$. Neither of these is a simple mental calculation. Substituting compatible numbers might give these calculations, ones that are easy to do mentally: $720 \div 90$ and $2,800 \div 40$.

57. The actual number falls between 3% of 3,000 (90 bulbs) and 3% of 4,000 (120 bulbs).

59. Responses will vary among students. A possible response is: overcooking, cooking for too long, produces dry meat. If Paul underestimates the cooking time (Paul says it will take 3 and a half hours when it will actually require 4 hours), then at the time for dinner, the turkey will not be ready. The guests will have to wait until the bird is done, but it will be moist when removed at the correct time. On the other hand, if he overestimates the cooking time, the bird will be done before the time for dinner and, if dinner is at the planned time, the bird may dry out. So, Paul should underestimate cooking time, but not by too much.

61. Responses will vary among students. A possible response is: It is better to underestimate contributions. Excess money can always be put to good use.

63. Answers will vary among students. Using rounding, we get the difference
    $1,600,000 - 1,000,000 = 600,000$ more teachers.

65. Responses will vary among students. Estimating the number of dogs in each category to the nearest million, we get in order, $25 + 19 + 6 + 2 + 1 = 53$ million dogs. The headline: **53 MILLION POOCHES IN U.S.**

67. Possible Answer. $60,645 \div 2,412$ is about $60,000 \div 2,000$. The Hearst Castle square footage is about 30 times the square footage in an average U.S. house.

69. Responses to this question depend on the insights developed by students as they have studied this section of the text.

71. Although answers will vary, students should recognize that computing technologies often are employed to obtain exact results of computation. But the user needs some method of determining if the device has been correctly used. Estimation techniques serve as a preliminary check on the correct use of calculators.

SECTION 3.3

1.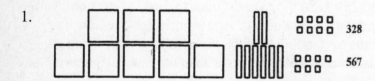

3. **Correct:** applying the distributive property of multiplication (in reverse) we have
   $(75 \times 38) + (75 \times 38) = (75 \times 38)(1 + 1) = 2(75 \times 38)$.

5. **Incorrect:** $6[2(3.5 + 8.6)] = 12(3.5) + 12(8.6)$. distributive

7. The statement IS correct, $0 = 0$, although the right expression is not an alternate form of the left expression.

9–11: Responses will vary in the manner in which the error is described and base-ten blocks are used to illustrate the errors. Possible responses include:

9. The error is in subtracting the smaller number from the larger ignoring which is the subtrahend and which is the minuend.

11. There are several errors. As above, although the ones place of the minuend was increased by ten units the tens place was not reduced by one ten. It appears that the hundreds place was reduced and when hundreds were subtracted the zero place holder in the difference was not inserted.

13. Ignore the parentheses and enter the numerals and operations in order. 28, +, 75, +, 134, –, 12, =. The result is **225.**

15. The keystroke sequence is: 789, –, 23, –, 45, –, 345, =. The result is **376.**

17. Jorge neglected to included the parentheses in his computation, so the minus sign was not properly distributed.

19. The difference in cost, D, is: D = (179 + 85) – (137 + 89) = **\$ 38**. Using a four function calculator the keystrokes are: 179, +, 85, –, 137, –, 89, =.

21. The answer can be obtained with the keystroke sequence: 1000, –, 179, –, 85, – ,68, –, 23, – ,47, –, 12, –, 115, –, 137, –, 89, =. The result is **\$ 245**.

23. Let one of the numbers be x, the other y. Then $x + y = 973$ and $x - y = 277$. We might add getting: $(x + y) + (x - y) = 973 + 277$. Applying commutativity and associativity we have: $x + x + (y - y) = 1250$. So $2x = 1250$ and one of the numbers is **625**. $y = 973 - 625$. So the second number is **348.**

25.
$$
\begin{aligned}
548 &= 5 \text{ hundreds} + 4 \text{ tens} + 8 \text{ ones} \\
+276 &= 2 \text{ hundreds} + 7 \text{ tens} + 6 \text{ ones} \\
\hline
&= 7 \text{ hundreds} + 11 \text{ tens} + 14 \text{ ones} \\
&= 7 \text{ hundreds} + 12 \text{ tens} + 4 \text{ ones} \\
&= 8 \text{ hundreds} + 2 \text{ tens} + 4 \text{ ones} \\
&= 824
\end{aligned}
$$

27.
$$
\begin{aligned}
1256 &= 1 \text{ thousand} + 2 \text{ hundreds} + 5 \text{ tens} + 6 \text{ ones} \\
+867 &= \phantom{1 \text{ thousand} + } 8 \text{ hundreds} + 6 \text{ tens} + 7 \text{ ones} \\
\hline
&= 1 \text{ thousand} + 10 \text{ hundreds} + 11 \text{ tens} + 13 \text{ ones} \\
&= 1 \text{ thousand} + 10 \text{ hundreds} + 12 \text{ tens} + 3 \text{ ones} \\
&= 1 \text{ thousand} + 11 \text{ hundreds} + 2 \text{ tens} + 3 \text{ ones} \\
&= 2 \text{ thousand} + 1 \text{ hundred} + 2 \text{ tens} + 3 \text{ ones} \\
&= 2123
\end{aligned}
$$

29. $136 = 1(10^2) + 3(10) + 6(1)$
 $+123 = 1(10^2) + 2(10) + 3(1)$
    $= 2(10^2) + 5(10) + 9(1)$
    $= 259$

31. $1393 = 1(10^3) + 3(10^2) + 9(10) + 3(1)$
 $+2736 = 2(10^3) + 7(10^2) + 3(10) + 6(1)$
    $= 3(10^3) + 10(10^2) + 12(10) + 9(1)$
    $= 3(10^3) + 11(10^2) + 2(10) + 9(1)$
    $= 4(10^3) + 1(10^2) + 2(10) + 9(1)$
    $= 4129$

33. (Say-give): $(84 - 0)(85 - 1)(90 - 5)(100 - 10)(120 - 20)(130 - 10)(135 - 5)$. The difference is 51.

35. (Say-give): $(32 - 0)(35 - 3)(55 - 20)$. The difference is 23.

37. **265**

39. **2688**.

37.
6 7	2	1
$7_1\,5_2$	$6_8\,7$	$2_3\,34$
7 5_7	$_1 7_5\,5$	$6_9\,54$
$4_1\,8_5$	$_1 7_2\,5$	$_1 7_6\,68$
5	$4_6\,8$	**1656**
	265	

39.
37 8	2	2
$98_1\,9_7$	3 $7_9\,8$	$3_5\,78$
$86_1\,4_1$	$9_1\,8_7\,9$	$_1 9_4\,89$
45 7_8	$8_1\,6_3\,4$	$_1 8_2\,64$
8	4 $5_8\,7$	$4_6\,57$
	88	**2688**

41. The addends are rewritten as: $175 + 352 = 100 + 75 + 300 + 52$.
The commutative property is applied: $100 + 75 + 300 + 52 = 75 + 52 + 100 + 300$.
Finally, the associative property is used: $(75 + 52) + (100 + 300) = 127 + 400 = 527$.
An alternative explanation would be the application of front-end estimation and exact adjustment.
Both approaches are valid.

43. The net result of 24 hours worth of work is 50 m of depth in the well. On first inspection then, it would seem that the well would be dug to a depth of 875 m on the eighteenth day. However, at the end of sixteenth day of digging, the well is 850 m deep. That night, rocks fall in the shaft so that it is only 800 m deep. During the seventeenth day the workers will pass the 875 m mark because they can dig 100 m per day.

45. By tracing out the routes we find 18 routes. Adding the mileage of some of these we get the following: ABCDEJK (1657), ACDEJK (1638), ABGFEJK (1339), ABGFIK (1256), ABGFIJK (1502), ABGHIK (1410), ABGHIJK (1656), ABCDEFGHIJK (2320).

47. Answers will vary. Another problem could be: What is the difference in distance between the shortest and longest routes?

49. Using mental mathematics and compensation:
$1349 - 875 = 1350 - 850 - 25 - 1 = 1300 - 800 - 25 - 1 = 500 - 25 - 1 = 475 - 1 = \textbf{474}$.

51. 38 tons $= 38$ tons $\times 2,000$ pounds per ton $= 76,000$ pounds. $76,000 - 8,000 = 68,000$. The King Kong in the original version weighed 68,000 pounds more than the King Kong in the 2005 version.

53. Answers will vary. Possible answer: The 1933 cost was about $600,000.
207 million $- \$600,000 = \$206,400,000$. The 2005 movie cost $206,400,000 more to take.

55. Responses depend upon interactions among students. A possible response could be modeled on question #3.

57. Responses will vary. A possible example is: in the above problem we found that $645 - 268 = 377$. We may check our subtraction by re-writing the subtraction relation as an addition equation: is it true that $645 = 377 + 268$? Since the answer is 'yes', the subtraction was correctly executed.

59. The third addend which produces a sum of 299 is 112. But the missing addend must be 2 digits. So the largest possible solution is **99** and the sum is 286.

SECTION 3.4

1.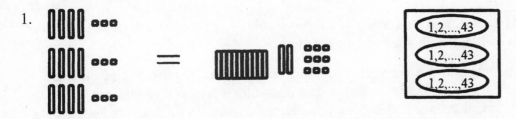

3. a. The partial products are 368 and 1380.
   b. The second partial product ends in 0 because the number of units in the product is completely in the first partial product. The 0 is "holding a place" until the partial products are added.
   c. Using rounding, an estimate is $40 \times 50 = 2,000$.

5. The basic computation was the multiplication of 12×22. The multiplicands were rewritten as sums, $22 \times 12 = (10 + 10 + 1 + 1) \times (10 + 1 + 1)$. Applying the distributive property we have:
$22 \times 12 = (10 + 10 + 1 + 1) \times (10 + 1 + 1) = (10 \times 10 + 10 \times 1 + 10 \times 1 + 10 \times 10 + 10 \times 1 + 10 \times 1 + 1 \times 10 + 1 \times 1 + 1 \times 1 + 1 \times 10 + 1 \times 1 + 1 \times 1) = 264$. The 2 10 by 10 products are the 2 large squares; the 6 10 by 1 products are the 6 rectangles; and 4 1 by 1 products are the 4 small squares.

7.
$$\begin{array}{r} 34 \\ \times\ 18 \\ \hline \end{array}$$

8×4 ----> **32**

8×30 ----> **240**

10×4 ----> **40**

10×30 ----> **300**

612

9. ince 10 fives is 50 and 20 fives is 100, the interval is **11–20**.

11. Since 25 fours is 100 and 20 fours is 80, the estimate is **21–30**.

13. Since 70 times 30 is 2100 and 70 times 2 is 140 and 2100 – 100 is less than 2065, the estimate is **61–70**.

15. Since 60 times 70 is 4200 and 60 times 4 is 240 and since 4200 + 240 > 4386, the estimate is **51–60**.

17. Estimates will vary. A possible estimate is: changing to compatible numbers we approximate $4357 \div 32$ with $4300 \div 43 = \mathbf{100}$. A calculator gives 136 full groups of 32 in 4357.

19. Keystrokes: 2208, ÷, 12, ÷, 23, =, **8**

21. Keystrokes: 15, ×, 12, ÷, 8 , +, 124, = **147**

23. Seventy five objects are to be equally shared among 5 sets. Thus the problem is: $Q = 75 \div 5$.

25. Answers will vary. Possible answers are given below.

There are a total of 145 photos from a wedding. If an album holds 6 photos on each page, how many pages are needed to hold all of the photos from the wedding?

27. The student is not combining place values to find the solution. When finding that 5×8 is 40, the student has written that product in the answer rather than recording 0 ones and placing the 4 tens above the 4 in 48. The same was done in the second problem; the result of 7×2 was recorded in the product rather than recording the ones and placing the tens above the 3 in 32.

29. First 45 is broken into addends, (40 + 5) and then the distributive property is applied.

31. An estimate is $80 \times 90 = \mathbf{7200}$.

33. An estimate is $70 \times 90 = \mathbf{6300}$.

31.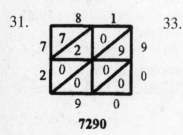

7290

33.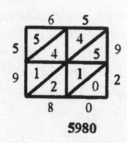

5980

35. Since I know that $32 \div 8 = 4$, I know that $32 = 4 \times 8$ or that there are 8 groups of 4 in 32 or 4 groups of 8. Finding $32 \div 4$ is finding how many groups of 4 are in 32. But I know that is 8. So $32 \div 4 = 8$. A check on division is to multiply the divisor and the quotient, then add back the remainder, if any, and compare this number to the dividend. If they are equal, then the division has been correctly performed.

37. $\$1,505,124 \div \$12,521 \approx 120.2$. The winner of the Daytona 500 won approximately 120 times more money than the last place finisher.

39. 26,389 miles $\div 76$ hours ≈ 347.2 miles per hour. He traveled about 347 miles each hour.

41. 45,000 feet $\div 2,500$ feet $= 18$. The Global Flyer was 18 times higher than the private plane.

43. Compute 20 times the yearly difference with the adjusted calculation
$20(40,000 - 25,000) = 20(15,000) = \textbf{\$300,000}$.

45. Three times $\$23,000$ is $\$69,000$ in year 5. Now, in year 1 the driver makes $\$23,000$, year 2 $\$23,000 + \$8500$, year 3 $\$23,000 + 17,000$, and in year 4 $\$23,000 + \$25,500 = \$48,500$. Thus in year 5 the driver must increase by the difference between $\$69,000$ and $\$48,500$: $\textbf{\$20,500}$.

47–49. Responses will vary. Students may use a pure guess and check approach although some may see the value of using approximate square roots as starting points.

47. Since the target is 50,000, begin with 200×200 because 200 squared is 40,000 and 300 squared is 90,000. Now try to narrow your possibility down by using a calculator $222 \times 222 = 49,284$. $222 \times 224 = 49,728$. $221 \times 226 = 49,946$. Students may continue until satisfied.

49. The largest product is $666 \times 666 = \textbf{443,556}$.

51. To save $\$1$ million in 35 years, saving the same amount each year, one must save $\$1,000,000 \div 35$ years $= \textbf{\$28,571}$ per year. If this is one fourth yearly earnings, one must earn $4(\$28,571) = \textbf{\$114,285}$ per year.

53. You want to obtain the sum: $75,000 + 200 + 400 + 800 + \cdots + 53,687,091,200 + 107,374,182,400$. This can, of course, be done by brute force. One might note that the sum is also represented by: $S = 75,000 + 200(2^0 + 2^1 + \cdots + 2^{29})$. This latter is a geometric series with first term 1, common ratio 2, and 30 terms. Some students may recall the sum of such a series is $1(1 - 2^{30})/1 - 2 = 1,073,741,823$. So the sum is approximately $75,000 + 200(1 \text{ billion})$ or about 200 billion dollars.

55. Responses will vary. We want the quotient of 486 and 15. We might estimate by first finding, mentally, $450 \div 15 = 30$. The remaining 36 can be separated into 2 groups of 15, giving 32 bottles per carton with 6 left over.

57. Answers will vary. Possible responses are:
Use repeated subtraction: enter 207, −, 23, =, = until the display shows either 0 or a number less than 23. Obtain the quotient by counting the number of times the = is pressed.

59. Responses will vary. A possible answer is: 328 pencils are to be given away, 6 at a time, to students. How many students will receive a group of 6 pencils?

61. Responses depend upon student interactions.

63. Responses will vary. A possible response is: the stacking algorithm is, in a sense, using inverse operations. First, because the dividend has 1 more place than the divisor, a trial multiplication by 10 is attempted. Since the difference between the dividend and 10 times the divisor is a whole number, the 10 becomes a partial quotient. Now 10 times the divisor is removed (subtracted) from the dividend and the process is repeated. Another 10 group of the divisor is determined and subtracted, this 10 groups added to the divisor. Since the dividend has been reduced to the same number of places as the division, 2 groups are removed leaving a number less than the divisor. The numbers of the groups of the divisor removed are added to obtain the quotient.

CHAPTER 3 REVIEW EXERCISES

1. 30, 80, 81, 82, 83, **84**.

3. 648, 548, **448**

5. $(5 \times 13) \times (20 \times 2) = (5 \times 20) \times (13 \times 2) = 100 \times 26 = \mathbf{2600}$

7. $8 \times 12 \times 5 = 8 \times (12 \times 5) = 8 \times 60 = \mathbf{480}$

9. $548 + 261 = 500 + 200 + 40 + 60 + 8 + 1 = 700 + 100 + 9 = \mathbf{809}$

11. $431 \times 2 = 2 \times (400 + 31) = 800 + 62 = \mathbf{862}$

13. $78 \times 2 = 80 \times 2 - 2 \times 2 = 160 - 4 = \mathbf{156}$

15. $253 + 58 = (253 - 1) + 58 + 1 = 252 + 58 + 1 = 310 + 1 = \mathbf{311}$

17. $41 - 27 = (41 + 3) - (27 + 3) = 44 - 30 = 14$

19. $210 - 88 = (210 + 2) - (88 + 2) = 212 - 90 = \mathbf{122}$

21. **5700**

23. **7000**

25. $665 + 243, 700 + 200 = 900$

27. $75 \times 15, 80 \times 10 = 800$

29. $3426 - 352, 3400 - 400 = 3000$

31. $33 \times 8, 30 \times 10 = 300$

33. $436 + 735 = 1100 + 70 = 1170$

35. $43 + 56 + 62 + 87 = 230 + 20 = 250$

37. $1243 + 1119 + 1228 + 1210, 4(1200) = 4800$

39. $2546 + 4337$, estimate: $2500 + 4500 = 7000$. Exact, **6883**

41. $540 - 308$, estimate: $540 - 300 = 240$. Exact: **232**.

43. $50 \times 30 = \mathbf{1500}$; $60 \times 40 = \mathbf{2400}$.

45. $450 \div 3 = \mathbf{150}$; $480 \div 3 = \mathbf{160}$

47.

49. Responses will vary. Responses may include the ideas: compatible numbers are numbers that are easy to compute with mentally. Different persons have different compatible numbers. Estimation is generally a mental process. For example, in adding groups of figures, $7 + 6 + 2 + 8 + 4 + 3 + 12 + 8$ a common practice is to scan for pairs that add to 10.

51. There are $12 \div 2$ possible pairs of jars: 4 choice for the first, 3 for the second: divided by 2 because the order of jars is not important. The sums are (in thousands): $154 + 25 = \mathbf{179}$, $154 + 750 = \mathbf{904}$, $154 + 3.5 = \mathbf{157.5}$, $25 + 750 = \mathbf{775}$, $25 + 3.5 = \mathbf{28.5}$, $750 + 3.5 = \mathbf{753.5}$. The greatest difference is $904 - 28.5 = \$875.5$ K

53. Students may attempt this by guess-check-revise. For example:
$5000 - $ (tennis vacation + clubs) $= 966$. So add the video equipment for 599. Now 407 must be returned. Continue by trading, for example the mountain bike for the video equipment, return 310, and including until satisfied with the amount to be returned.

55. $3,164 \text{ lbs.} \div 39 \text{ lbs.} \approx 81.1$. The car weighs about 81 times as much as the pocket bikes.

57. NY to LA $2 \times \$144 = \$288$. DC to SD $2 \times \$99 = \$198$. $\$288 - \$198 = \$90$. A round trip ticket from New York to Los Angeles costs \$90 more than a round trip from Washington D.C. to San Diego.

59. $3 \times (2 \times \$125) = \$750$. Three round-trip tickets for Denver to Washington D.C. will cost \$750.

61. Responses will vary. A possibility is: the common algorithms all involve regrouping either into equivalent larger groups or into equivalent smaller groups. The size of groups is determined by place value.

SECTION 4.1

1. Divide 79 by 13. If the remainder is 0, 13 is a factor of 79. In this case, 79 divided by 13 leaves a remainder of 1, so 13 is not a factor of 79.

3. The Factor <u>Test</u> Theorem assures us that the largest number we need to test to determine the factors of 729 is $\sqrt{729}$, or 27.

5. By division, we see that the possible factors are 7, 11, 13, 77, 91, 143, and by multiplication we see that the possible multiples are: 1001, 2002, 3003, etc.

7. 221 divided by 17 is 13 with no remainder, so 17 is a factor of 221. $\sqrt{221} \approx 14.8$, so try factors up through 14. 13 is a factor of 221.

9. Because $6 \times 3 = 18$. 18 is a **multiple** of 6.

11. Since $1 \cdot a = a$ for any natural number a, 1 is a **factor** of any natural number.

13. Since if $c = a \cdot b$, a and b non-zero whole numbers, c is a non-zero multiple of a and a non-zero multiple of b, a non-zero **multiple** of a number is always greater than or equal to the number.

15. Since the cube of a number, a, is $a \times a \times a = a \times a^2$, a number is a **factor** of its cube.

17. Divisibility by 2, 3, 4, 5, 6, should be done mentally. The methods are:
    a. Even numbers are divisible by 2. Generally these are recognized by inspection. As a rule, however, the number is divisible by 2 if the right-most digit is divisible by 2.
    b. The rule is: a number is divisible by three if the sum of its digits is divisible by three. Note that this rule can be sequentially applied until a number that is, or is not, divisible by 3 can be determined by inspection. For example: Is 95132695684392 divisible by 3? The sum of its digits is 72. Is 72 divisible by 3? The sum of its digits is 9 which is divisible by 3. So 72 is divisible by 3 as is the original number.
    c. The rule for divisibility by 4 is: if the number formed by the last 2 digits is divisible by 4, then the number is divisible by 4. Recognition of numbers that are multiples of 4 permit this to be done by inspection.
    d. A number is divisible by 5 if the last digit is 0 or 5.
    e. A number is divisible by 6 if it is divisible by 2 and by 3.

19. Since 1, 2, 3, 6, and 9 are factors of 18, any number divisible by 18 would also be divisible by 1, 2, 3, 6, 9.

21. A number is divisible by 3 if the sum of its digits is divisible by 3: that is, the sum of the digits is a multiple of 3.

    $x\,7\,y\,z$: $x + y + z$ could $= 5$ to make the sum of the digits a multiple of 3. So the number could be **2712**.

23. A number is divisible by 6 if and only if it is divisible by 3 and by 2. The numbers 8370, 8376 represent the only ways to supply the missing digit so the numbers are divisible by 2 (end in an even number) and also by 3 (sum of the digits is divisible by 3.)

25. The following numbers are divisible by both 2 and 3: 56178, 56478, 56778.

27. A natural number is divisible by 9 if and only if the sum of its digits is divisible by 9.

    Let 73_5_ be represented by $73x5y$. Using the same reasoning as in the previous problem, $7 + 3 + x + 5 + y$ must be a multiple of 9. A possible solution is 73053.

29. (a) Divisible by 2, 3, 4, 5, 6, 8, 9, 10, 11. (b) Divisible by none of the numbers. (c) Divisible by 3, 5, 7. (d) Divisible by all of the numbers.

31. (a) $9 - (5 + 4) = 0$, and $11 \mid 0$, so $11 \mid 495$. (b) $(5 + 2 + 6) - (9 + 0 + 4) = 0$, and $11 \mid 0$, so $11 \mid 642,059$. (c) $(0 + 9 + 8 + 7) - (3 + 4 + 6 + 2 + 3) = 6$. $11 \nmid 6$ so $11 \nmid 372,869,403$.

33. The proper factors of 28 are 1, 2, 4, 7, 14 and $1 + 2 + 4 + 7 + 14 = 28$, so 28 is a perfect number.

35. The proper factors of 24 are 1, 2, 3, 4, 6, 8, 12. The sum of these factors is 36, so 24 is an abundant number.

37. a. $9 = 3 \times 3$ is a square number with factors 1, 3, 9: an odd number of factors.
    b. $16 = 4 \times 4$ and is even; $25 = 5 \times 5$ and is odd.
    c. $36 = 6 \times 6$ and has proper factors 1, 2, 3, 4, 6, 9, 12, 18 with a sum of 55. So it is even, square, and abundant.

39. Yes. She could use the divisibility test for 4. Since the last two digits of 232 are divisible by 4, 232 is divisible by 4.

41. The 84 spaces could be laid out in rectangular arrays with the lengths and widths of each rectangle pairs of factors of 84. These are: **1 by 84, 2 by 42, 3 by 28, 4 by 21, 6 by 14, and 7 by 12**.

43. Use a calculator to form a several numbers *abcdabcd*, for example 16001600, 14451445, 19701970. Divide by 73. The results are integer. Thus one may conjecture any repeated 4 digit number is divisible by 73 and thus any repeated birth year is divisible by 73.

45. **Yes, any number divisible by 10 is divisible by 5**. Suppose that N is divisible by 10. Then N = 10 × Y, Y some natural number. Now N = 10 × Y = 5 × (2 × Y). Thus 5 is a factor of N and N ÷ 5 = 2 × Y. However, **a number divisible by 5 is not necessarily divisible by 10**. Consider the numbers 5, 15, 25, 35, and so on, numbers ending in 5. They are all divisible by 5 but not by 10.

47. Some numbers with exactly 3 factors are 4 (1,2,4), 9 (1,3,9), 25 (1,5,25), 49 (1,7,49), 121 (1,11,21). Since 1 is a factor for all natural numbers, 1 will be 1 of the 3 factors in the set we are concerned with. The second factor is a number that has only 2 factors (primes), one of which is 1. The third factor is the number itself, the square of the second factor. Thus **numbers with only 3 factors are squares of primes**.

49. a. **False**. Let n be 6, a be 3, and b be 4. It is true that $6 \mid 3 \cdot 4$ but $6 \nmid 3$ and $6 \nmid 4$.
    b. **False**. Consider that 6 divides $(7 + 5)$ but 6 does not divide 5 or 7.
    c. **True**. As an example, 3 divides 6 and 3 divides 9. 3 also divides $6 + 9$. In general, if n divides a and n divides b, then $a = nx$ and $b = ny$. Now, $a + b = nx + ny = n(x + y)$. Thus n is a factor of $a + b$ so n divides $a + b$.
    d. **True**. 4 divides 8 and 8 divides 16. 4 also divides 16. In general, if n divides a and a divides b then $a = nx$ and $b = ay$. So $b = nxy$. Since n is a factor of b, n divides b.
    e. **False**. Consider that 6 divides $(7 + 5)$ but 6 does not divide 5 or 7.
    f. **True**. Since $n \mid a$ and $n \mid b$ with $a > b$, $a = un$ and $b = vn$, u and v natural numbers, $u > v$. So $(a - b) = (u - v)n = w(n)$, w a natural number. Thus $n \mid (a - b)$ by definition of 'divides'.

51. The proper factors of 496 are: 1, 2, 248, 4, 124, 8, 62, 16, 31. The sum of these factors is 496. So 496 is a perfect number.

53.

numbers	operation	result	numbers	operation	result
even, even	+	even	even, even	$\times$	even
even, odd	+	odd	even, odd	$\times$	even
odd, odd	+	even	odd, odd	$\times$	odd

55. a. $1 \times 9 + 3 \times 6 + 2(0) + (-1)1 + (-3)2 + (-2)3 = 14$. Since 14 is divisible by 7, so is 321,069.
    b. Answers will vary. It might be easier, since the mental math required involves smaller numbers.

57. a. The 3, 5, and 7 digit palindromes are not divisble by 11, while the 2, 4, and 6 digit ones are, verified 57.   as follows: $3,223 = 11 \times 293$, $467,764 = 11 \times 42,524$.
    b. All palindromes with even number of digits are divisible by 11.

59. Let N be the number we are seeking. If N divided by seven leaves a remainder of 0, then N must be a multiple of 7. When N is divided by 2, 3, 4, and 5 the remainder is 1. Thus 2, 3, 4, 5 are factors of N – 1. Since a number with a factor of 4 also has a factor if 2, N – 1 is a multiple of $3 \cdot 4 \cdot 5 = 60$. The smallest such multiple is 60 itself. So N itself is a (multiple of 60) + 1 that is divisible by 7. Multiples of 60, plus 1, are 61, 121, 181, 241, 301, ... Now, 301 divided by 7 = 43, so 301 is a solution. Trial shows that others are 721, 1141, and 1561. Note that $301 = 5(60) + 1$, $721 = 12(60) + 1$, $1141 = 19(60) + 1$, and $1561 = 26(60) + 1$. This pattern suggests that the solutions are of the form N = $(5 + 7n)$ 60 + 1, n = 0, 1, 2, 3, 4, ...

61. a. Begin by adding the factors from largest to smallest, checking the sum after each addition. The sum of 78 and 52 is 130. The sum of 130 and 39, 169, is greater than 156. So 156 is abundant.
    b. Janie might apply some estimation techniques. Replace all the factors with the next higher number of 10's: 10, 10, 20, 30, 40, 100. The sum of these, which is an overestimate of the exact sum, is seen to be 210, less than 273. Thus 273 is deficient.

63. Answer: (b) Since the *contrapositive* of the above italicized statement is true, if the sum of the remainders of the addends had not been equal to the remainder from the sum, you could have concluded that the calculation had not been done correctly. However, since the *converse* of the italicized statement is not necessarily true, you can only *suspect* that the calculation in the example was *probably* done correctly.

65. The sum in e appeared to be correct by the test, but in reality was incorrect. It was a case where the same digits appeared as in the correct answer, but two of them were transposed.

67. a.   All numbers, such as 247,247, of the form *abc,abc* are divisible by 7, 11, and 13.
    b.   Since $7 \times 11 \times 13 = 1001$, and 1001 times any three digit number *abc* is *abc, abc*, then we know that 7, 11, and 13 are always factors of a number of the form *abc, abc*.

69. Responses will vary. Possible responses include: $4 \mid 32$. But 32 can be expressed as $(20 + 12)$ and $4 \mid 12$. So we have a situation in which $n \mid (a + b)$ and $n \mid b$. But $4 \mid 20$. So we also have $n \mid a$. Thus $4 \mid (12 + 20)$ is an example of a situation in which since $n \mid (a + b)$ and $n \mid b$, $n \mid a$.

71. The statement is completed: **"4 divides *r*, the last two digits of the number."** By the **Divisibility of Sums Theorem**, if 4 divides $100q$ and also divides *r*, then 4 will divide $100q + r$ which is the original number.

73. Let $a = 3$ and $b = 12$. Then $a \mid b$. Let $c = 5$. Then $3 \mid 12 \bullet 5$. Whenever $a \mid b$, then $ax = b$, *x* a whole number. Multiplying both sides by *c*, we have $axc = bc$; $a(cx) = bc$. Since the product of two whole numbers is a whole number, by definition $a \mid bc$.

75. If the palindrome is *abba*, for example, it is divisible by 11 provided $(a + b) - (b + a)$ is divisible by 11. But $11 \mid 0$, so *abba* is divisible by 11. A similar argument could be given for any even digit palindrome.

77. Solutions will vary. A possible solution is: Solve by reducing the set of 720 possible 6-digit numbers by successively applying the conditions of the problem. Since the number formed by the first 5 digits is divisible by 5 and there are no zeros in the number, digit 5 must be 5. The conditions that the number formed by the first 2 digits is divisible by 2 and there are no duplicate numbers restricts the first 2 digits to the set $\{12, 14, 16, 24, 26, 32, 34, 36, 42, 46, 62, 64\}$. Now applying the condition that the number formed by the first 3 digits is divisible by 3 and reserving 5 for the 5th digit, we obtain the set of potential first 3-digits: $\{123, 126, 162, 243, 261, 321, 324, 342, 423, 621\}$. To add the fourth digit we use the 'divisibility by 4 test'. The set of first-5-digit possibilities is: $\{12365, 12645, 16245, 24365, 32165, 42365\}$. Finally, the entire number must be divisible by 6. This entails being divisible by both 2 and 3. Since the sum of the digits 1 through 6 is 21, any number formed by all six of these digits is divisible by 3. So the final digit must be 2, 4, or 6 to ensure divisibility by 6. Thus we have **123654** and **321654** as the possible license plate sequences.

79. Responses will vary. One solution is: let the house number be *XYZ*. Now, since the house number is even, then *Z* must be 0, 2, 4, 6, or 8. But since the product *XYZ* is non-zero, we can conclude that none of the digits is 0. Finally, because the house number is assumed to be unambiguous and we are given no information that serves to distinguish *X* from *Y*, we can assume that *X* and *Y* are equal. If $Z = 2$, $X + Y = 12$ and $XY = 36$. Because $X = Y$, both *X* and *Y* are 6 and a possible address is 662. If $Z = 8$, *X* and *Y* are 3 and a possible address is 338. Since Cindy refers to an oldest (singular) child, the address is **338**.

81. Because the hint says a perfect number is involved, try \$6. It is true that \$1 + \$2 + \$3 = \$1 × \$2 × \$3 = \$6. So, the first three items could have cost \$1, \$2, and \$3. To find the price of the fourth item, we can solve \$6x = \$6 + x because we know that adding the new price to the total gives the same result as multiplying the total by the new price. Solving this equation for x gives $6x - x = \$6$; $x - (1/6)x = 1$; $(5/6)x = 1$; $x = 6/5$. So the fourth item costs \$1.20.

83. The proper factors of 120 are: 1, 2, 3, 4, 5, 6, 8, 10, 12, 15, 20, 24, 30, 40, 60. The sum of these is 240, 2 times 120. The proper factors of 672 are: 1, 2, 3, 4, 6, 7, 8, 12, 14, 16, 21, 24, 28, 32, 42, 48, 56, 84, 96, 112, 168, 224, 336. The sum of these numbers is 1344 which is 2 times 672.

SECTION 4.2

1. By definition a prime number has 2 and only 2 distinct factors, 1 and the number itself.

3. A composite number has at least 3 distinct factors.

5. Primes are c) 13 and f) 2, since they have exactly two factors.
   a. $2|16$, so 16 isn't prime.
   b.   By definition of Prime Number, 1 is not prime.
   d. $3|81$, so 81 is not prime.
   e.   0, by definition is not prime

7. **Answer for Exercise 7**

	2*	3^	4*	5°	6*^	7'	8*	9^	10*°
11	12*^	13	14*'	15°	16*	17	18^*	19	20*°
21^'	22*	23	24*^	25°	26*	27^	28*'	29	30*^°
31	32*	33^	34*	35°'	36*^	37	38*	39^	40*°
41	42*^'	43	44*	45°	46*	47	48*^	49'	50*°
51^	52*	53	54*^	55°	56*'	57^	58*	59	60*^°
61	62*	63^'	64*	65°	66*^	67	68*	69^	70*°'
71	72*^	73	74*	75°	76*	77'	78*^	79	80*°
81^	82*	83	84*^'	85°	86*	87^	88*	89	90*^°
91'	92*	93^	94*	95°	96*^	97	98*'	99^	100*°

*Multiples of 2   °Multiples of 5
^Multiples of 3   'Multiples of 7

The numbers that are not multiples of 2, 3, 5, or 7 are the prime numbers less than 100: 11, 13, 17, 19, 23, 29, 31, 37, 41, 43, 47, 53, 59, 61, 67, 71, 73, 79, 83, 89, 97

9. a. $1260 \div 2 = 630$, $630 \div 2 = 315$, $315 \div 3 = 105$, $105 \div 3 = 35$, $35 \div 5 = 7$.
        Prime factorization = **2 × 2 × 3 × 3 × 5 × 7**.
   b. $3250 \div 2 = 1625$, $1625 \div 5 = 325$, $325 \div 5 = 65$, $65 \div 5 = 13$.
        Prime factorization = **2 × 5 × 5 × 5 × 13**.
   c. $1105 \div 5 = 221$, $221 \div 13 = 17$. Prime factorization = **5 × 13 × 17**.
   d. $5940 \div 2 = 2970$, $2970 \div 2 = 1485$, $1485 \div 3 = 495$, $495 \div 3 = 165$, $165 \div 3 = 55$, $55 \div 5 = 11$.
        Prime factorization = **2 × 2 × 3 × 3 × 3 × 5 × 11**.

11. Responses will vary among students. A possibility is: use the Factor Test Theorem to determine the possible natural number factors. The square root of 437 is between 20 and 21 so only natural numbers less that 21 need be tested. Divisors of 437 are 1, 19, 23, 437. Since there are more than 2 distinct factors, **437 is not** prime. The square root of 541 is between 23 and 24 so only natural numbers less than 24 need be tested for divisibility. Trial divisions on a calculator show no natural numbers less than 24 divide 541. Thus 541 has no factors other than 1 and 541. So **541 is prime**.

13. The factors of 42 are: 1, 2, 3, 6, 7, 14, 21, 42. The factors of 28 are 1, 2, 4, 7, 14, 28. The common factors are 1, 2, 7, 14. The GCF is **14**.

15. a. $36 = 2^2 \times 3^2$ $48 = 2^4 \times 3$. Using the greatest exponent that occurs in both numbers' prime factorizations, GCF $= 2^2 \times 3 = 12$.
    b. $51 = 3 \times 17$ $21 = 3 \times 7$. Using the greatest exponents that occur in both numbers' prime factorization GCF $= 3$.
    c. $60 = 2^2 \times 3 \times 5$ $225 = 3^2 \times 5^2$. Using the greatest exponents that ocuur in both numbers' prime factorization, GCF $= 3 \times 5$ or 15.

17. Answers will vary. Example: $156 = 2^2 \times 3 \times 13$ $504 = 2^3 \times 3^2 \times 7$. Using the greatest exponents that occur in both numbers' prime factorization, the GCF is $2^2 \times 3$, or 12.

19. a. $888 \div 259 = 3\, R\, 111$; $259 \div 111 = 2\, R\, 37$; $111 \div 37 = 3\, R\, 0$. So GCF $=$ **37**.
    b. $308 \div 84 = 3\, R\, 56$; $84 \div 56 = 1\, R\, 28$; $56 \div 28 = 2\, R\, 0$. GCF $=$ **28**.
    c. $7560 \div 1232 = 6\, R\, 168$; $1232 \div 168 = 7\, R\, 56$; $168 \div 56 = 3\, R\, 0$. GCF $=$ **56**.

21.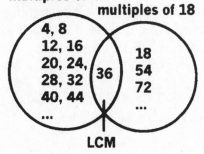

23. a. $27 = 3^3$
        $35 = 5 \times 7$
        So, LCM $(27, 35) = 3^3 \times 5 \times 7 = 945$
    b. $60 = 2^2 \times 3 \times 5$
        $28 = 2^2 \times 7$
        so, LCM $(60, 28) = 2^2 \times 3 \times 5 \times 7 = 420$
    c. $36 = 2^2 \times 3^2$
        $56 = 2^3 \times 7$
        So, LCM $(36, 56) = 2^3 \times 3^2 \times 7 = 504$

25. Answers will vary. Sample answer:
   $42 = 2 \times 3 \times 7$
   $70 = 2 \times 5 \times 7$
   $\text{GCF}\,(42, 70) = 2 \times 7 = 14$
   $\text{LCM}\,(42, 70) = 2 \times 3 \times 5 \times 7 = 210$

27. a.   3 is the only factor of 3 other than 1, and using the divisibility by 3 test, we see that 3 is not a factor of 894, 367, 235. So GCF of the two numbers is 1.
   b.   Using the divisibility tests, neither 2 nor 4 is a factor of 631, 457, 247, so the GCF of the two numbers is 1.

29. If the numbers are a and b, $\text{GCF}(a,b) \times \text{LCM}(a,b) = a \times b$. So $6 \times 36 = 12 \times b$, $b = (6 \times 36) \div 12 = \textbf{18}$.

31. 2000, 2002, 2004, 2006, 2008, and 2010 are divisible by 2, not prime. 2001 and 2007 are divisible by 3 (sum of its digits is divisible by 3), not prime. 2005 is divisible by 5, not prime. 2009 is divisible by 7, not prime. So 2003 is the only prime number year in the first 10 years of the 21st century. None of the primes up to the square root of 2003 are divisors of 2003.

33. $30 = 2 \times 3 \times 5$, $42 = 2 \times 3 \times 7$. so, $\text{GCF}\,(30, 42) = 6$, and the largest small-group size is 6.

35. The solution to this exercise is the LCM of 24 and 36. The GCF is 12.
   So the LCM is $(24 \times 36)/12 = \textbf{72 hrs}$.

37. Times that are multiples of 12 minutes represent whole numbers of laps for the faster cycle and times that are multiples of 15 minutes represents whole numbers of laps for the slower cycle. To pass the starting line at the same time during the race each bike would have to travel a whole number of laps. The first time that this will happen is the LCM of the individual lap times. So: $12 = 2 \times 2 \times 3$. $15 = 3 \times 5$. The LCM is $2 \times 2 \times 3 \times 5 = 60$. The two cycles will cross the line together in **60 minutes**. The faster cycle will have gone 5 laps, the slower will have gone 4 laps.

39. Substituting and calculating we get $(n, n^2 - n + 11)$:(1, 11); (2, 13); (3, 17); (4, 23); (5, 31); (6, 41); (7, 53); (8, 67); (9, 83); (10, 101); (11, 121). **All are prime but 121**, which occurs when $n = 11$.

41. Trial gives **2, 5, 17, and 37** as primes less than 100 of the form $n^2 + 1$

43. Responses will vary. A possible response is: the formula $p = 6n - 1$, for n a natural number, generates the odd numbers that are 1 less than consecutive multiples of 6. The first few of these are: 5, 11, 17, 23, 29, 35, 41, 47, 53, 59, 65, 71, 77, 83, 89, 95, 101. Of these 17 numbers 13 are prime. But, 5 of the first 5 are prime, 9 of the first 10, and 13 of the first 17. The percentage of primes is decreasing. Since there are an infinite number of numbers in the set determined by $6n - 1$, the percentage well might drop below 50% and then continue to drop.

45. a.   The primes greater than 2 are: 3, 5, 7, 11, 13, 17, 19, 23, 29, 31, 37, 41, 43, 47, 53, 59, 61, 67, 71, 73, 83, 89, 97. The ones digits of primes greater than 2 are odd numbers.
   b.   Except for 2, all primes are odd. So the possible ones-digit sequences for three consecutive odd numbers are: 1, 3, 5; 3, 5, 7; 5, 7, 9; 7, 9, 1; and 9, 1, 3. No number ending in 5 that is greater

than 5 is a prime because a number ending in five has 5 as a factor. So only 2 of the 5 sequences is even possible. Thus I conjecture that groups of 3 consecutive primes are rare.

   c. Others pairs of reversal primes are 13 and 31, 17 and 71, 37 and 73, 107 and 701, and 11 which is its own reversal prime.

47. Responses will vary. Students' responses should include the concept that computer systems are limited in terms of the size of the numbers that they can handle. Computers are finite digital machines.

49. By the GCF-LCM Theorem, if the GCF is 1, since GCF $\times$ LCM = the product of the 2 numbers, then $1 \times$ LCM is the product and hence the LCM is the **product of the two numbers**.

51. Results will vary among students. Possible responses are:
   a. The prime 61 can be represented as $4 \times 15 + 1$. Now, $61 = 25 + 36$, the sum of 2 square numbers.
   b. Let 12 be the number. Then 24 is the double. Between 12 and 24 we find the primes 13, 17, 19, and 23.
   c. Let p be 2. Then $2^p - 1 = 3$, a prime. Now, $(2^{2-1})(2^2 - 1) = 6$. The proper factors of 6, 1, 2, and 3 add to 6. Thus 6 is a perfect number.
   d. $16 \div 2 = 8$, $8 \div 2 = 4$, $4 \div 2 = 2$, and finally, $2 \div 2 = 1$. On the other hand, for odd numbers, consider $7 : 3(7) + 1 = 22$; $22 \div 2 = 11$; $3 (11) + 1 = 34$; $34 \div 2 = 17$; $3(17) + 1 = 52$; $52 \div 2 = 26$; $26 \div 2 = 13$; $3(13) + 1 = 40$; $40 \div 2 = 20$; $20 \div 2 = 10$; $10 \div 2 = 5$; $3(5) + 1 = 16$; $16 \div 2 = 8$; $8 \div 2 = 4$; $4 \div 2 = 2$; $2 \div 2 = 1$.

53. Responses will vary. A possible response is: three consecutive whole numbers may be represented by $N, N + 1, N + 2$. Of these 3 numbers at least 1 must be even. But because an even number is divisible by 2, it cannot be prime (unless it IS 2. But in that case if $2 = N + 1$, then $N = 1$ and is not prime. If $N = 2$, then $N + 2$ is even and not prime). Thus there are not 3 consecutive primes.

55. Let $m = 12$, $n = 9$. GCF $(12, 9) = 3$. $m + n = 21$, LCM$(12, 9) = 36$. GCF$(21, 36) =$ GCF$(3 \times 7, 3 \times 3 \times 2 \times 2) = 3$. Let $m = 25$, $n = 35$. GCF$(25, 35) = 5$. $m + n = 60$. LCM $(35, 25) = 175$. GCF$(60, 175) =$ GCF$(2 \times 2 \times 3 \times 5, 5 \times 5 \times 7) = 5$. The 2 examples **do not** prove the generalization true. There may be pairs of numbers, m and n, for which it is not true. The proof, if it exists, must be deductive.

57. The procedure involves determining how many times each prime serves as a factor of each number. The greatest number of these appear as factors in the LCM. The trial divisions terminate when the final quotient is 1 because that 1 will be the final factor. As an example consider the problem of determining the LCM of 24, 28, and 45.

2	24	28	45
2	12	14	45
2	6	7	45
3	3	7	45
3	1	7	15
5	1	7	5
7	1	7	1
	1	1	1

So LCM = $2 \times 2 \times 2 \times 3 \times 3 \times 5 \times 7 =$ **2520**.

$24 \div 2 = 12$; $12 \div 2 = 6$, $6 \div 2 = 3$. Thus 2 appears as a factor of 24 3 times. $28 \div 2 = 14$, $14 \div 2 = 7$. 2 is a factor of 28 twice. So far, then, the LCM contains the greatest of these, 3 2's. Now, 2 does not divide 45. So 2^3 is a factor of the LCM. Now we move on to 3. $3 \div 3 = 1$ so 3 is a factor once in 24. 3 does not divide 7 so 3 is not a factor of 28. $45 \div 3 = 15$ and $15 \div 3 = 5$. So the greatest number of 3's is 2 and the LCM contains 3^2 as a factor. The remaining numbers are prime and appear once in the LCM. So the LCM = $\mathbf{2^3 \times 3^2 \times 5 \times 7 = 2520}$.

59. The factors of 36, obtained by using natural numbers no greater than the square root of 36, as trial divisors, are: 1, 36, 2, 18, 3, 12, 4, 9, and 6. There are 9 factors. One could also use the method of 36. $36 = 2^2 \times 3^2$. Increasing the exponents by 1 and multiplying gives 9 as the product. $9 - 2 = 7$, the number of different polygons that can be constructed.

61. Solutions will vary. A possible solution is: let B represent the number of large boxes, b the number of small boxes in each large box, p the number of packages in each small box, and s the number of sticks of gum in each package. So $(B)(b)(p)(s) = 39039$. The last 3 of these numbers are primes less than 15. The prime factorization of 39039 is $(13)(13)(11)(3)(7)$. The possible numbers of large boxes are the products of any 2 of the prime factors provided such product has 2 digits. So the possibilities are: 39, 91, 33, 77, and 21.

63. The cycles will next coincide after a number of days has passed that is the LCM of the lengths of each of the cycles. 23 is a prime number. the prime factorization of 28 is $2 \times 2 \times 7$. The prime factorization of 33 is 3×11. Thus the LCM is $23 \times 2 \times 2 \times 7 \times 3 \times 11 =$ **21,252** days. In this time 924 physical cycles will have been completed, 759 emotional cycles, and 644 intellectual cycles.

65.

```
        i. All numbers            ii. Every 2ⁿᵈ out     iii. Every 3ʳᵈ out     iv. Every 7ᵗʰ out
  1  2  3  4  5  6  7  8  9  10    1  3  5  7  9          1  3     7  9          1  3     7  9
 11 12 13 14 15 16 17 18 19  20   11 13 15 17 19            13 15     19           13 15
 21 22 23 24 25 26 27 28 29  30   21 23 25 27 29         21    25 27           21    25 27
 31 32 33 34 35 36 37 38 39  40   31 33 35 37 39         31 33     37 39        31 33     37
 41 42 43 44 45 46 47 48 49  50   41 43 45 47 49            43 45     49           43 45     49
 51 52 53 54 55 56 57 58 59  60   51 53 55 57 59         51    55 57           51    55 57
 61 62 63 64 65 66 67 68 69  70   61 63 65 67 69         61 63     67 69           63    67 69
 71 72 73 74 75 76 77 78 79  80   71 73 75 77 79            73 75     79           73 75     79
 81 82 83 84 85 86 87 88 89  90   81 83 85 87 89         81    85 87                 85 87
 91 92 93 94 95 96 97 98 99 100   91 93 95 97 99         91 93     97 99        91 93     97 99
```

```
    v. Every 9ᵗʰ out        Every 13th out         Every 15ᵗʰ out          Every 21ˢᵗ out
  1  3     7  9          1  3     7  9          1  3     7  9          1  3     7  9
     13 15                  13 15                  13 15                  13 15
 21    25               21    25               21    25               21    25
 31 33    37            31 33    37            31 33    37            31 33    37
    43 45    49            43       49            43       49            43       49
 51    55               51    55               51                     51
    63    67 69            63    67 69            63    67 69            63    67 69
 73 75    79            73 75    79            73 75    79            73 75    79
    85 87                  85 87                  85 87                     87
 93    97 99            93       99            93       99            93       99
```

a. After removing the single 21ˢᵗ number, 85, we are left with the **23** lucky numbers less than 100:
   l, 3, 7, 9, 13, 15, 21, 25, 31, 33, 37, 43, 49, 51, 63, 67, 69, 73, 75, 79, 87, 93, 99. Of these
   23 lucky numbers **9 are prime**, **13 are composite,** and 1, the number 1, is neither.

b. There are **7 pairs of twin primes**: 1/3, 3/5, 5/7, 11/13, 17/19, 41/43, 59/61, 71/73 and **7 pairs of twin lucky numbers**: 7/9, 13/15, 31/33, 49/51, 67/69, 73/75.

c. Both primes and lucky numbers have 7 pairs differing by 4 in the sets less than 100.

d. Because 4 = 1 + 3, 6 = 3 + 3, 8 = 1 + 7, 10 = 3 + 7, 12 = 3 + 9, 14 = 7 + 7, 16 = 3 + 13,
   18 = 9 + 9, and 20 = 7 + 13 we may conjecture that any even number greater than 2 can be
   written as the sum of 2 lucky numbers.

67. For $n = 0$, $P_n = 3$; for $n = 1$ we get 5; for $n = 2$ we get 17; for $n = 3$, we get 257; for $n = 4$ we get
   65,537 – all primes. For $n = 5$, the expression for P_n gives 4,294,967,297 = 641(6,700,417),
   a composite.

CHAPTER 4 REVIEW EXERCISES

1. To determine the factors of a number n, divide n by the natural numbers less than or equal to the
   square root of the number. If the remainder is zero, then both the divisor and quotient are factors
   of n. Applying this strategy we get:
   a. The factors of 18 are: **1, 18, 2, 9, 3, 6**.
   e. The factors of 32 are: **1, 32, 2, 16, 4, 8**.

f.   The factors of 48 are: **1, 48, 2, 24, 3, 16, 4, 12, 6, 8**.
g.   The factors of 105 are: **1, 105, 3, 35, 5, 21, 7, 15**.

3.   To determine the factors of a number *n*, divide *n* by the natural numbers less than or equal to the square root of the number. If the remainder is zero, then both the divisor and quotient are factors of *n*. Applying this strategy we get:
a.   The factors of 91 are: **1, 7, 13, 91**.
b.   The factors of 143 are: **1, 143, 11, 13**.
c.   The factors of 663 are: **1, 663, 3, 221, 13, 51, 17, 39**.
d.   The factors of 299 are: **1, 299, 13, 23**.

5.   a.   The numbers **1436, 4674, and 5580** are divisible by 2 because the last digit is even.
b.   The numbers **987, 4674, and 5580** are divisible by 3 because the sums of the digits is divisible by 3.
c.   The numbers **1436 and 5580** are divisible by 4 because the number formed by the last 2 digits is divisible by 4.
d.   Only **5580** is divisible by 5 because the last digit is either 5 or 0.
e.   The numbers **4674 and 5580** are divisible by 6 because they are divisible by 2 and by 3.

7.   11 is a prime because it has exactly two factors, the number 11 itself, and 1. 10 has four different factors, 1, 2, 5, 10.

9.   a.
$$\begin{array}{ll} 90 & 420 \\ 2 \times 45 & 2 \times 210 \\ 2 \times 3 \times 15 & 2 \times 5 \times 42 \\ 2 \times 3 \times 3 \times 5 & 2 \times 5 \times 2 \times 21 \\ & 2 \times 5 \times 2 \times 3 \times 7 \end{array}$$
b. $90 \div 2 = 45; 45 \div 3 = 15; 15 \div 3 = 5$. $420 \div 2 = 210; 210 \div 5 = 42; 42 \div 2 = 21; 21 \div 3 = 7$.

11. a.   The set of factors of 48 is {1, 48, 2, 24, 3, 16, 4, 12, 6, 8} The set of factors of 108 is {1, 108, 2, 54, 3, 36, 4, 27, 6, 18, 9, 12}. The intersection of these sets is {1, 2, 3, 4, 6, 12} and the largest element of the intersection is the GCF, **12**
b.   The set of prime factors of 48 is {2, 2, 2, 2, 3} and the set of prime factors of 108 is {2, 2, 3, 3, 3,}. The intersection of these sets is {2, 2, 3}. So the GCF is $2 \times 2 \times 3 = $ **12**.
c. $108 \div 48 = 2 \, R \, 12. \, 48 \div 12 = 4 \, R \, 0$. Thus the GCF is **12**.

13. a.   No. They have a common factor of 13.
b.   Yes, they have no common factors other than 1.

15. Arguments may vary. A possibility is: If an even number other than 2 were prime then it would have no factors other than itself and 1. But all even numbers are divisible by 2. Thus 2 is a factor of all even numbers. So no even number other than 2 is prime.

17. Arguments will vary. A possible argument is: If *N* is to be divisible by 2, 3, 4, 5, 6, and 7, each of these must be factors of *N*. But if 6 is a factor of *N*, so are 2 and 3. So keep 2 and 3 and drop 6. Similarly, since 4 is a factor, so is 2. So keep 4 and drop 2. So we have as factors 3, 4, 5, and 7. The product of these is **420**.

19. a.   We must find the LCM of 90 and 240, convert this time to hours, and add that number of hours to 12:00 noon. $90 = 2 \times 3 \times 3 \times 5$. $240 = 2 \times 2 \times 2 \times 2 \times 3 \times 5$. So the LCM is $2 \times 2 \times 2 \times 2 \times 3 \times 3 \times 5 = 720$ min. 720 min = 12 hours. The sirens will sound simultaneously at midnight.

   b.   Since $720 \div 90 = 8$, A will sound again 8 times for a total of 9. B will sound $(720 \div 240) + 1$, 4 times.

   c.   Siren C could be set to sound every 80 minutes.

21. a.   The square plots must have both length and width equal to the GCF of 560 and 528. The prime factorizations are: $560 = 2 \times 2 \times 2 \times 2 \times 5 \times 7$. and $528 = 2 \times 2 \times 2 \times 2 \times 3 \times 11$. The GCF = 16. The largest research plots are **16 m × 16 m**.

   b.   The total area divided by the area of a research plot $[(560 \times 528) \div (16 \times 16)] = \mathbf{1155}$.

   c.   Smaller plots could have an edge length that is a factor of 16: 1, 2, 4, or **8**.

23.   A chart of individual tests might be:

Potential divisor, d	d divides dividend, D, if:
2	D ends in 0, 2, 4, 6, or 8
3	The sum of the digits of D is divisible by 3
4	The number formed by the last 2 digits of D is divisible by 4
5	D ends in 0 or 5
6	D is divisible by both 2 and 3
7	The number formed by subtracting twice the last digit from the number formed by all but the last digit is divisible by 7
8	The number formed by the last 3 digits is divisible by 8
9	The sum of the digits is divisible by 9
10	The number ends in 0

Some of the concepts included in the responses might include: the divisibility test by two is, as described, essentially a way of identifying even numbers. All even numbers have a factor of 2. If the last digit of a number is 0, then the number is the product of the number formed by all the digits except the last and 10. Thus 10 is a factor and the number is divisible by 10. If a number is written in expanded form it is seen that each place is divisible by 10, and thus by 5. But the last (units) position is divisible by 5 iff it is a 5 or 0. Divisibility by both 2 and 3 ensures divisibility by 6. And if a number is not divisible by 6 it does not have both 2 and 3 as factors and is not divisible by both 2 and 3.

Responses depend upon group interactions of the students for pro and con discussions of the various methods.

SECTION 5.1

1. a.   +10, write a check for $10: −10 .
   b.   −6, score 6 points in a game: +6 .
   c.   −8, temperature 8 degrees above zero, +8 .
   d.   +500, live in a town 500 ft below sea level: −500

3. The opposite of a negative integer is a **positive** integer.

5. The absolute value of a nonzero integer is always a **positive** integer.

7. (a)

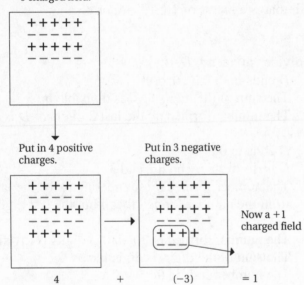

(b)

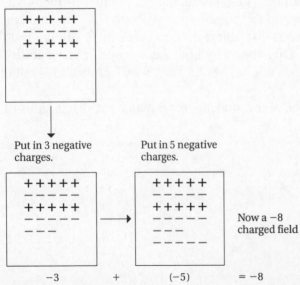

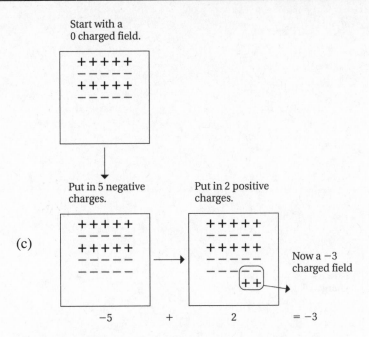

9. Responses will vary. Possible responses are:
   a. 11 is represented by a set of 11 black chips. The opposite of 5 is represented by a set of 5 red chips. The union of the two sets is formed and each red chip is paired with a black chip. There are 6 unpaired black chips. Thus the sum of 11 and –5 is 6.
   b. Begin at 0 on the number line and move 7 units to the left to represent the opposite of 7. Then move 9 units to the right to represent the addition of 9. The final position on the number line is 2. So –7 + 9 = 2.
   c. Form the union between a set of 6 red chips and a set of 8 black chips. Pair red and black chips and count the unpaired chips. Since there are 2 unpaired black chips, the result of adding the opposite of 6 to 8 is 2.

11. a. Associative property of addition: $(5 + 4) + – 4 = 5 + (4 + – 4)$
    b. Additive inverse property: $(4 + –4) = 0$
    c. Additive identity property: $5 + 0 = 5$.

13. (a)

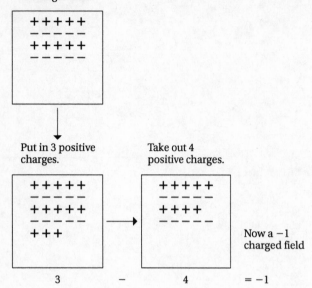

(b)

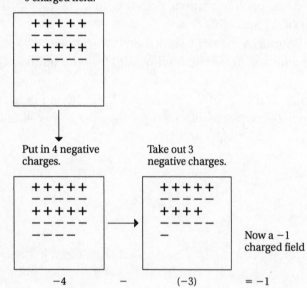

(c)   Start with a
      0 charged field.

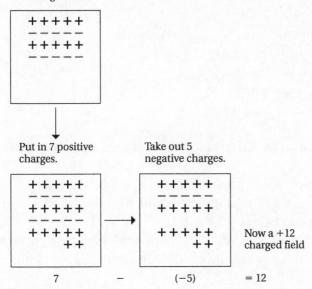

7   —   (−5)   = 12

15. a.   23 Answers will vary.
    b.   869 Answers will vary.
    c.   −2 . Answers will vary.

17. a.   −6, since − 6 adds to − 3 to give − 9.
    b.   −21, since − 21 adds to 8 to give −13.
    c.   22, since 22 adds to −7 to give 15.

19. a. So |−8| is = **8**.   b. −|8| = **−8**.   c. −|- 8| = **−8**.   d. |−(- 8)| = |8| = **8**.   e. |6 − 9 | = |-3| = **3**.

21. $\left|-6\right|-\left|6\right| = 6-6 = \mathbf{0}$

23. $-\left|-14\right|-(-12) = -14+12 = -2$

25. $-\left\|28+(-17)\right|+(-6)\right| = -\left|11+(-6)\right| = -\left|5\right| = -5$

27. Conceptually, LOSS + GAIN = NET. Losses are represented by negative integers and gains by positive integers. So the decrease in the price of the stock is represented by LOSS = −5, the increase by GAIN = 9. So we have **−5 + 9 = NET**. So the NET or TOTAL GAIN for the 2 days is **$4**.

29. a.   The order, largest to smallest, is: **21, 8, 0, −4, −6, −9, −12, −15**.
    b.   The order, largest to smallest, is: **23, 18, 17, 9, −2, −12, −15, −18**.
    c.   The order, largest to smallest, is: **−13,570, −13,999, −14,000**.

31. Since −5 s to the right of −9 on the number line, it follows that −5 > −9.

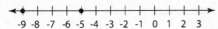

33. To solve the equation a – (–459) = 9236, first change it to the equivalent equation $a + 459 = 9236$. So $a = 9236 – 459 = 8777.

35. Approaches and calculator key sequences will vary. Possibilities include:
    a. The key sequence: 643, $+/–, +, 754, +/–, =,$ gives $–1397$.
    b. The key sequence: 6732, $–, 7845, = +/–,$ gives $14,577$.
    c. The key sequence: 1357, $+/–, +, 3429, =,$ gives $2,072$.
    d. The key sequence: 8435, $+/–, –, 9568, +/–, =,$ gives 1133.

37. a. The pattern could be generated by successive additions of 4. So the pattern might be –8, –4, 0, **4, 8, 12**.
    b. The pattern could be generated by successive subtractions of 7: 15, 8, 1, **–6, –13, –20**.
    c. One way to generate the first 3 numbers after 5 is to subtract 15, add 20, subtract 30. Continuing the pattern, add 40, subtract 50, add 60, to get: –20 + 40 = 20, 20 – 50 = –30, –30 + 60 = 30. So the sequence becomes: 5, –10, 10, –20, **15, –30, 20, –40**. To generate this pattern, look at every other number. The positive numbers are successive multiples of 5 and the negative numbers are successive multiples of –10.
    d. The first 5 numbers in the pattern could be generated by beginning with 1, then successively adding –3, 4, –5, 6, ... to each successive number. Doing this produces: 1, 1 + –3 = –2, –2 + 4 = 2, 2 + –5 = –3, –3 + 6 = 3, 3 + –7 = **–4**, –4 + 8 = **4**, 4 + –9 = **–5**.

39. a. A week from now would be represented by the positive integer **7**.
    b. Yesterday would be represented by the negative integer **–1**.
    c. A week ago would be represented by **–7**.
    d. The day after tomorrow would be represented by 1 + 1 = **2**.
    e. A month from now could be represented by **30**.

41. The additive inverse of an integer is another integer such that the sum of the 2 integers is 0
    a. $–k + k = 0.$ b. $(j + k) + –(j + k) = 0.$ c. $j + –j = 0.$ d. $(j – k) + – (j – k) = 0 = –j + k.$

43. Let $n = 7$. Since $7 > 0$, the definition says that $|7| = 7$, which is true. Now let $n = –4$. Since $–4 < 0$, the definition says that $|–4| = – (–4) = 4$, which is true.

45. Methods of solution will vary. Possibilities include:
    a. Since 1 must be added to –6 to get –5, $x = 1$.
    b. Adding 4 to –9 we find that $n = –5$.
    c. Adding 8 to 12 we find that $y = 20$.
    d. Subtracting – 4 from –4 + 12, we find $z = –4 + 12 – (–4) = 12$.
    e. Since $t – (–9) = 2$ is equivalent to $t + 9 = 2$ we can add –9 to each side of the equation. So $t = –7$.
    f. Subtracting –4 from 7 + –4 gives 7 + (–4) – (–4) = r. So $r = 7$.

47.

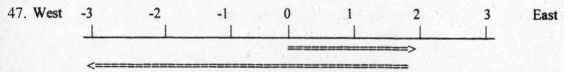

The final position of the car is three miles west of the station.

49. One might begin this exercise by ordering the numbers: −61, −49, −29, −21, 39, 45, 56.
    a. **−49 + 45** = −4.
    b. **−61 + (−49)** = −110.
    c. **39 + 56** = 95.
    d. **39 − (−61)** = 100.
    e. **−21 − (−29)** = 8.
    f. **39 − 45** = −6.

51. Estimates and methods will vary. Possibilities include:
    a. Estimate by **rounding**: 700 − 500 = **200**.
    b. All of the values **cluster** around −500. So −500 − 500 − 500 − 500 − 500 = 5(−500) = **−2500**.
    c. Using **compatible numbers** and **breaking apart** the result is approximated by:
        −300 − 50 + 50 + 1200 = **900**.
    d. **Front-end estimation** gives the approximation −800,000 + 300,000 = **−500,000**.
    e. **Rounding** to estimate gives −400 − 800 + 200 + 1000 − 700 = **−700**.

53. To complete the magic square first find the missing addend in the diagonal (7) the missing addend in
    row 2, and so on. The square is:

−2	3	−7	4
0	−3	7	−6
5	−8	2	−1
−5	6	−4	1

55. Using the numbers −3, −2, −1, 0, 1, 2, 3 the following triplets sum to 2: (−3, 3, 2), (−2, 3, 1),
    (−1, 1, 2), (−1, 0, 3). These can be arranged in a triangle the following ways, so that each side has a
    sum of 2:

1	3
−2 2	−3 0
3 0 −1	2 1 −1

57. Consider that (−a + −b) + (a + b) = 0 and also that −(a + b) + (a + b) = 0. Since the additive inverse
    of (a + b) is unique, (−a + −b) = −(a + b).

59. The story will vary among students. The facts are: The home team scored on its first possession with
    80 net yards of offense. After a successful conversion the home team led 7 − 0. The visitors received
    the ball on the kickoff and returned it 24 yards. On their first possession the visitors punted after a
    net gain of 6 yards. The visitors netted **30** yards including the kick-off return.

61. After 2 rounds Drivefar is 3 strokes above Puttgood. Puttgood picks up an additional 7 strokes on
    7 holes. So with 11 holes yet to be scored, Puttgood is 10 strokes ahead of Drivefar. But Drivefar can
    gain 1 stroke on each of the eleven holes. He has the potential to win by 1 stroke.

63. Responses will vary. One possibility is: the arrow model is similar to the colored counter model in
    that there are a variety of representations of any integer because an opposite pair, red/black counters
    or left/right arrows represent 0. So as 6 could be represented by 6 black counters or 8 black and 2 red
    or 10 black and 4 red it could also be represented by 6 unit arrows to the right or 8 arrows to the
    right and 2 to the left or 10 to the right and 4 to the left.

65. a.   In order to remove two red counters, 2 red/black pairs must be added. After removing the 2 reds, the two black counters remain. The net effect is that of adding two black counters. So to subtract two reds, one might add the opposite of two reds, that is two black counters.

    b.   Responses will vary. From the definition of subtraction, $a - b$ is the unique integer that adds to b to give a. From the adding opposites theorem, $a + (-b)$ is also the unique integer that add to b to give a. Hence, $a - b = a + (-b)$.

67. Responses will vary. One possibility is: The elevator is similar to the number line. Designate some floor as 0, and agree that going upwards is moving in the positive direction and downwards in the negative direction.

69. a.   Responses will vary. One possibility is: yes, Arnauld raises a valid question. –1 is less than 1 because there is a positive integer, 2 such that –1 + 2 = 1. So –1/1 is indeed "a smaller to a greater" and 1/–1 is "a greater to a smaller". So how can a smaller, perhaps interpreted as a part, be related to the whole in the same manner as the whole is related to the part?

    b.   There are 2 values for x for which the sentence $x^2 + 2x - 8 = 0$ is true. These are 2 and –4. Bhaskara was telling his students to ignore the negative solution for "it is inadequate," perhaps to represent a quantity deemed important in that place at that time.

71. Responses will vary. They may include: $I = P \cup N \cup R$; $I = W \cup N$; $E = P \cap N$; $P \cap R = E$; $N \cap R = E$.

SECTION 5.2

1. a. $0(-5) = 0$ b. $-1(-5) = 5$ c. $-2(-5) = 10$ d. $-3(-5) = 15$

3. a.

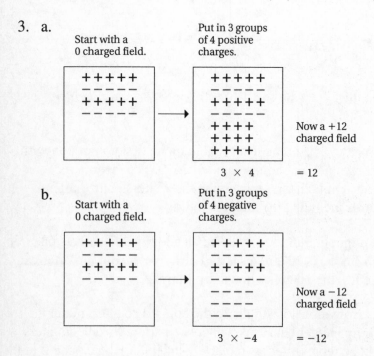

    b.

c.

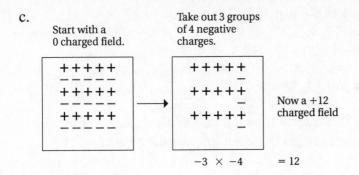

Start with a
0 charged field.

Take out 3 groups
of 4 negative
charges.

Now a +12
charged field

$-3 \times -4 \qquad = 12$

5. Answers will vary.
    a. −48
    b. 452,666
    c. −140
    d. −375

7. $|-8| \, (-13) = 8(-13) = \textbf{−104}$.

9. $-|8| \, (-8) \, (-8) = -8 \, (-8) \, (-8) = \textbf{−512}$.

11. $-|-56|(14 + (-9)) = -56 \, (5) = \textbf{−280}$.

13. a. −6786   b. 29,392   c. −98,901

15. **384**

17. **15,625**

19. **positive**

21. The associative property for multiplication (a b) c = a(b c).
    a. Let $a = -2$, $b = 3$, $c = -4$: $((-2)(3)) \, (-4) = 24$; $-2 \, ((3)(-4)) = -2(-12) = 24$.
    b. Let $a = 5$, $b = -3$, $c = -2$: $((5)(-3))(-2) = -15(-2) = 30$; $5((-3)(-2)) = 5(6) = 30$.
    c. Let $a = -6$, $b = -5$, $c = -4$: $((-6)(-5))(-4) = -120$; $(-6)((-5)(-4)) = -6(20) = -120$.

23. a. 9 times what is −72? **−8**.
    b. −7 times what is −42? **6**.
    c. −9 times what is 63? **−7**.

25. $-6(12 + (-7)) = -6(5) = \textbf{−30}$.

27. $-6((-54) \div (-9)) = -6(6) = \textbf{−36}$.

29. $(-7)(-9) + 7(-9) + -9(7) = 63 \, -63 - 63 = \textbf{−63}$.

31. $(53 + (-5)) \div (12 - (-4)) = 48 \div 16 = \textbf{3}$.

33. The average change is $-$ \$12 M/4 months = **$-$\$3M/month**.

35. a.   Score = $16(5) + 8(4) + 8(2) + 3(-2) + 11(-4) + 6(-5) = $ **48**.
    b.   Average negative score is $[3(-2)] + 11(-4) + 6(-5)] \div 20 = $ **-4**.
    c.   Average positive score is $[16(5) + 8(4) + 8(2)] \div 32 = $ **4**.

37. Answers will vary. Possibilities are: 1. Enter 92, enter -23, and push the divide key. 2. Enter 92, then enter 23 and push the repeat subtract 4 times to get 0. Recognize that the answer will be negative. In either case, $92 \div (-23) = -4$.

39. a. $-4n = 20$, $n = 20 \div -4 = $ **-5**
    b. $n/3 = -6$, $n = -6(3) = $ **-18**
    c.   no integers makes the equation true.
    d.   All integers make the equation true.
    e. $n^2 = 9$, $n = \sqrt{9}$, $n = $ **3** or $n = $ **-3**

41. Responses will vary. Possibilities include:
    a. $-2(-y) = 2y$: For all pairs of integers a and b, $(-a)(-b) = ab$ by the property of opposites.
    b.   Since $-(2x - 4) = (-1)(2x - 4)$ and $(-1)(2x - 4) = (-1)(2x) - (-1)(4)$ by the distributive property and $(-1)(2x) - (-1)(4) = -2x + 4$ by the property of opposites, we have $-(2x - 4) = -2x + 4$.
    c. $-1(ab) = -(ab)$ by the property of opposites.

43. Positive integer exponents represent the number of times a number is to be used as a factor in multiplication
    a. $5^3 = 5 \times 5 \times 5 = $ **125**   b. $(-2)^2 = (-2) \times (-2) = $ **4**   c. $(-5)^3 = (-5) \times (-5) \times (-5) = $ **-125**
    d. $(-3)^4 \div (-3)^3 = (-3)(-3)(-3)(-3)/(-3)(-3)(-3) = $ **-3**

45.

X	Pos	Neg	0
Pos	Pos	Neg	0
Neg	Neg	Pos	0
0	0	0	0

47 a.   A possible pattern is based on successive additions of $-6, 8, -12, 14, -18, 20, -24, 26, \ldots$ The absolute values of the additions alternate between 2 and 4 and the signs alternate. So, beginning with 2, we have the pattern: $2, (2 - 6 = -4), (-4 + 8 = 4), (4 - 12 = -8), (-8 + 14 = 6),$ $(6 - 18 = -12), (-12 + 20 = 8), (8 - 24 = -16), (-16 + 26 = $ **10**$), (10 - 30 = $ **-20**$), (-20 + 32 = $ **12**$)$.
   b.   A possibility is a sequence beginning with 1 and each succeeding term is obtained by multiplying by -2. This produces: $1, -2, 4, -8, 16, -32,$ **$64, -128, 256$**.
   c.   A possibility for this pattern is successive multiplications by $-3, 2, -6, 18, -54, 162,$ **$-486, 1458,$** **-4374**.
   d.   A possibility is to successively add numbers which have absolute values that are determined by multiplying 7 by successive powers of 2 and then appending alternating signs to get the successive addends beginning with $a-$. So that addends are $-7, 14, -28, 56, -112, 224, -448,$ 896. So, successively adding these numbers to each term beginning with 2 we get: $2, -5, 9, -19,$ $37, -75,$ **$149, -299, 597$**.

49. Responses will vary. A possible argument is: $-a$ is, by definition, the opposite of a. Thus $-a + a = 0$. Now consider the expression $[a(-1) + a]$: $a(-1) + a = a(-1) + a(1)$ [multiplicative identity] $= a(-1 + 1)$ [distributive property] $= a(0)$ [definition of additive inverse] $= 0$ [zero property]. Thus both $-a$ and $a(-1)$ are additive inverses of a. But the additive inverse of an integer is unique. So $-a = a(-1)$.

51. The argument assumes that $-1 \times 1 = -1$, and that $-1/-1 = 1$

53. Responses will vary. Student responses may include: Whereas integer multiplication is associative, integer division is not: $(24 \times 6) \times 2 = 24 \times (6 \times 2) = 288$ but $(24 \div 6) \div 2 = 2$ but $24 \div (6 \div 2) = 8$. Multiplication is commutative, division is not. $12 \times 2 = 2 \times 12$ but $12 \div 2$ is not equal to $2 \div 12$.

55. a. Additive identity.
    b. Distributive property.
    c. Additive identity.
    d. The additive identity element in the set of integers is a unique element. Since both 0 and $a \cdot 0$ add to $a \cdot 1$ to yield $a \cdot 1$, they must be different representations of the same element and thus be equal.

57. The total of the overages and underages is: $8(-3) + 6(-5) + 9(4) + 2(-7) = -32$. So the total weight is 32 lb under 25(50). Total $= 25(50) - 32 =$ **1218 lbs**.

59. Responses will vary. A possible response is: Negative integers represent bills, positive integers represent cash. Cash is necessary to pay bills. More cash is required to pay bigger bills than is required to pay smaller bills. When one says that a \$50 bill is larger than a \$25 bill that person is actually referring to the cash required to pay the bill, not suggesting that -50 is a larger number than -25. The absolute value of -50 is greater than the absolute value of -25 but -50 itself is less than -25.

61. Responses will vary. One possibility is: A Korean mathematician may have formed a set from the union of 4 sets of 3 black rods. Counting the number of rods in this set he arrived at 12 black rods which is represented as -12.

63. The responses to this question depend upon group interactions among students. In their discussions they may address the following: multiplication can be performed by repeated addition and division by repeated subtraction. Multiplication and addition are commutative and associative but subtraction and division are neither. The integers are closed under addition, multiplication and subtraction but not under division. Multiplication and division with operands of the same sign produce positive results.

CHAPTER 5 REVIEW EXERCISES

1. a. **True**.
    b. **False. The sum of a positive and a negative integer may be positive, negative, or 0**.
    c. **False**. The absolute value of 0 is 0 and 0 is not positive.
    d. **False. The product of 2 negative integers is always positive**.
    e. **False. The integer 0 is neither positive nor negative**.

3. a. **–9**   b. **13**   c. **–27**   d. **–24**

5. a. The number 2 can be represented by a set of counters in which there are 2 more black than red. So represent 2 with 10 black and 8 red counters. The process of subtraction is represented physically by removing counters, black for positive numbers, red for negative. So remove 5 red counters. We are left with a set with 10 black and 3 red counters. This set represents the number 7.
    b. The field [+ + + – + – + – + – + –] represents $2 + 0 + 0 + 0 + 0 = 2$. –(–5) means take away 5 –'s from the field. Doing this leaves [+ + + + + + +] which represents 7.
    c. Begin at the position on the number line that represents 2. Now, do the opposite (the subtraction sign) of moving left 5 units (the number –5). So move right 5 units. The final position, 7, represents $2 – (–5)$.
    d. The opposite of –5 is 5. Executing subtraction by adding the opposite, then, yields $2 + 5 = 7$.

7. The larger of two integers is farther to the right on a number line.

9. a. Since the second factor is –4 we are moving sets of 4 red counters. Because the absolute value of the first factor is 3 we are moving 3 sets of 4 red counters. Since the sign of the first factor is negative we will remove 3 sets of 4 red counters from a set that originally contained equal numbers of red and black counters. This set now contains 12 more black than red counters. So the product is 12.
    b. Start with a 0 charged field. Take out 3 groups of 4 negative charges. The result is +12 charged field. $–3 (–4) = 12$.
    c. On the number line we will do the opposite of moving left 4 spaces 3 times beginning at 0. This means that we will move right 4 units 3 times. The final position is 12.
    d. $3(–4) = –12$, $2(–4) = –8$, $1(–4) = –4$, $0(–4) = 0$ $–1(–4) = 4$ $–2(–4) = 8$ $–3(–4) = 12$.

11. a. Distributive property of multiplication over addition.
    b. Associative property of multiplication.
    c. Additive inverse property.
    d. Distributive property.
    e. For all non-zero integers b, $b \div b = 1$.
    f. The associative property of multiplication and the property of opposites.

13. Order the months:
    a. Nov Dec Jan Feb Mar Apr May Jun July Aug Sep Oct Nov Dec
                      –5   –4   –3   –2   –1   0   1   2   3   4   5   6
        Jan = –5, Mar = –3, May = –1, Aug = 2, Oct = 4, Dec = 6.
    b. Counting to count backwards on the above diagram, –7 is associated with **November**.

15 a. Associative property of addition.
    b. Additive inverse property.
    c. Additive identity property.

17. Let H represent the record high temperature and let L represent the record low. $H – L = 192$. Also $H – 4 = –L$. So we have, from the first equation, $H = 192 + L$. Thus $192 + L – 4 = –L$; $188 + L = –L$; $2L = –188$; $L = \mathbf{–94}$. From the first relation, $H = 192 + L = 192 + (–94) = \mathbf{98}$.

19. Responses will vary. The facts are that the test showed that the sample of 50 bags was a total of 53 oz light. This is about 1 oz per bag. A bag is sold at a weight of 50 lb, 800 oz. Essentially, the company is getting the income for 800 bags when selling 799 bags worth of fertilizer.

21. The tables are:

+	P	N
P	P	S
N	S	N

−	P	N
P	S	P
N	N	S

×	P	N
P	P	N
N	N	P

÷	P	N
P	P	N
N	N	P

The discussion of the value and limitations will vary among students. Responses may include the ideas that the tables are valuable as summary devices but are limited by the *S* entries.

SECTION 6.1

1.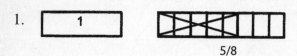

   5/8

3.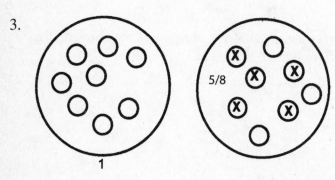

5.

7. Letting the hundreds square represent 1, the thousands cube represents 10 because 10 hundreds equals 1000. The tens stick represents 0.1 because 1/10 of a hundred is 10. Finally, the units cube represents 0.01 because 1/100 of one hundred is 1.

9. a. $4 \times 1/2 = 2$ b. $-18 \times -1/18 = 1$ c. $1/2 \times 6 = 3$ d. $386/1000 \times 1000 = 386$

11.

	Fraction form	Decimal form
a.	160/25	**6.4**
b.	**386/1000**	0.386
c.	**2/3**	0.666....
d.	4/11	**0.3636...**

13.
![number line from 0 to 1 with points 0.15, 3/10, 1/3, 0.8, 4/5, 1]

15. (a)
   0.28 is a little more than $\frac{1}{4}$.

   (b)
   0.72 is a little less than $\frac{3}{4}$.

   (c)
   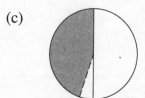
   0.45 is a little less than $\frac{1}{2}$.

(d)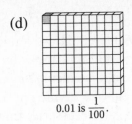

0.01 is $\dfrac{1}{100}$.

(e)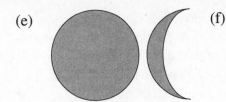

1.18 is a little more than 1.

(f)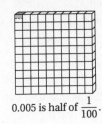

0.005 is half of $\dfrac{1}{100}$.

17. Explanations of reasoning may vary. Possibilities include:
   a. 0.28 is about **1/4** because 0.28 is about 0.25 and 0.25 = 25/100 = 1/4.
   b. 3.125 is about **3 1/10** because 0.125 rounds to .1.
   c. 0.03939.. is about **4/100 = 2/50 = 1/25** because 0.039 rounds to 0.04.
   d. .99... **is 1**. Let $n = 0.999....$ Then $10n = 9.999....$ and $10n - n = 9n = 9.999... - 0.999... = 9$. Thus $n = 1$.

19. Responses will vary. A possibility is: 0.3 is the decimal representation of 3/10 and 0.30 is the decimal representation of 30/100. But applying the Fundamental Theorem of Fractions to 3/10 by multiplying both numerator and denominator by 10, we get 3/10 = 30/100. Since the fractions represent the same number the equivalent decimal representations also represent the same number.

21. Responses will vary. One possibility is to represent the project with the number 1. Then if n students show up and commit to the project for the month, each student should do about $1/n$ of the work. If the project is anticipated to take h hours, the each student should plan on working $(1/n)(h)$ hours.

23. Responses will vary. Possible responses include: In both the rational and the whole numbers, the other numbers are based on 1. In the whole number system, the other numbers are multiples of 1, in the rational number system the other numbers are parts of 1. 1 is the multiplicative identity in both systems. In the system of rationals there are infinitely many representation of 1 (2/2, 3/3, etc.).

25. Responses will vary. The facts are that 1/13 of the women on one campus were polled and 48/1000 responded that they would prefer to attend all female mathematics classes. On the other campus 1/5 of the women students were polled and 48/1000 of those polled said yes.

27. Responses will vary. One possibility is: Adding 0's to the right of a decimal digit is adding 0 times a unitary part of some power of 10. But 0 times any number is 0. Thus when appending 0's one is actually adding 0's and 0 is the additive identity.

29.

Fraction	Decimal representation
1/7	**0.142857142.**
2/7	**0.285714285**
3/7	**0.428571428**
4/7	**0.571428571.**
5/7	**0.714285714**
6/7	**0.857142857**
7/7	**1.**

a. b. c. d. Responses may vary. The decimal representations of 2/7, 3/7, 4/7, 5/7, 7/7, and 7/7 are not the corresponding multiples of the decimal display for 1/7 if that display is reentered. However, if the display is the result of dividing 1 by 7 and that result is directly used in the multiplications by 1, 2, etc. the corresponding display are matched. Even though rounding will occur, the calculator maintains numbers to a greater accuracy than it displays. Moreover the quotient of 1 divided by 7 is an infinitely repeating decimal. When a number is entered from the keyboard it is necessarily finite.

31. Responses will vary. A possibility is: The Stevin notation is akin to our expanded notation. Stevin's representation of 325/1000 is (3 × one tenth unit) + (2 × one hundredth unit) + (5 × one thousandth unit). It is as if the circled numbers are negative exponents for powers of 10.

33. a.   The women's wage, w, is 3/4 of the men's wage, m: $w = (3/4)m$. Now, $w + (1/4)w = (5/4)w$. This is a 25% increase in pay. But $(5/4)w = (5/4)(3/4)m = (15/16)m$. This is still less than the men's wage.

   b.   Suppose that both the men's and women's wages are daily and based on an 8 hour day. Let w and m represent these daily wages. If m and w stay constant but the women's work day is reduced to 6 hours, the hourly wage for women becomes $w/6$ and the men's hourly wage is $m/8$. But w is $(3/4)m$. So the women's hourly wage can be represented $(3/4)m \div 6 = m/8$. So the second strategy equalizes the hourly wages. But the gross earnings of women will still be 3/4 of those of men.

   c.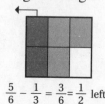

$ABCD$ = men
$DEFG$ = women

Needed for equality = $\frac{1}{4}$ $ABCD$
but $\frac{1}{3}$ $DEFG$

SECTION 6.2

1. (a) The large rectangle = 1.

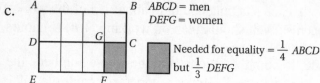

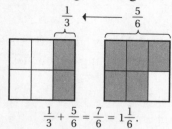

$\frac{1}{3} + \frac{5}{6} = \frac{7}{6} = 1\frac{1}{6}$.

(b) The large rectangle = 1.

$\frac{5}{6} - \frac{1}{3} = \frac{3}{6} = \frac{1}{2}$ left

(c) The large rectangle = 1.

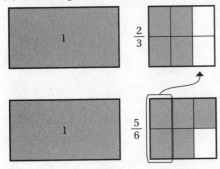

$$1\frac{2}{3} + 1\frac{5}{6} = 1\frac{4}{6} + 1\frac{5}{6} = 2\frac{9}{6} = 3\frac{3}{6} = 3\frac{1}{2}$$

(d) The large rectangle = 1.

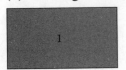 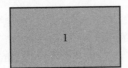

$$2\frac{2}{3} - 1\frac{1}{2} = 2\frac{4}{6} - 1\frac{3}{6} = 1\frac{1}{6} \text{ left}$$

3. a. $\dfrac{21}{4}$ b. $\dfrac{11}{3}$ c. $\dfrac{77}{10}$ d. $\dfrac{-7}{2}$

5. Answers will vary. One possibility is:

a. 1/3 + 3/4 + 2/3 = 1/3 + 2/3 + 3/4 = (1/3 + 2/3) + 3/4 = 1 + 3/4 = 1 3/4.
b. 5/6 + 3/8 + 4/6 = 5/6 + 4/6 + 3/8 = (5/6 + 4/6) + 3/8 = 9/6 + 3/8 = 3/2 + 3/8 = 12/8 + 3/8
    = 15/8 = 1 7/8.

7. Answers will vary. One possibility is to add 2 1/3 and 3 2/3 to get 6 and then subtract 1 2/3.
    Subtracting 1 gives 5 and then subtracting 2/3 gives 4 1/3. Another possibility is to start with 3 2/3
    and subtract 1 2/3, giving 2. Then add 2 1/3 for a total of 4 1/3. A third possibility is to combine the
    whole numbers (2 − 1 + 3) to get 4 and then combine the fractions, canceling the +2/3 and −2/3 and
    leaving 1/3. The final answer is again 4 1/3.

9. a. $9\frac{1}{3} + \frac{4}{5} = 9\frac{5}{15} + \frac{12}{15} = 9\frac{17}{15} = 10\frac{2}{15}$

b. $\frac{1}{2} + \frac{3}{4} + \frac{3}{5} = \frac{10}{20} + \frac{15}{20} + \frac{12}{20} = \frac{37}{20} = 1\frac{17}{20}$

c. $7\frac{1}{12} - 3\frac{1}{2} = 7\frac{1}{12} - 3\frac{6}{12} = \frac{85}{12} - \frac{42}{12} = \frac{43}{12} = 3\frac{7}{12}$

d. $\frac{7}{12} - \frac{3}{8} = \frac{14}{24} - \frac{9}{24} = \frac{5}{24}$

11. a.   1 because 3/5 is close to 1/2.
    b.   1/2 because 9 1/6 is close to 9 and 8 2/3 is close to 8 1/2.
    c.   1 1/2 because 1/4 + 1 3/4 is 2 and 2 – 1/2 = 1 1/2.
    d.   2 1/2 because 8 1/5 – 8 1/6 is close to 0, 12 1/3 is close to 12 and 14 3/4 is about 2 1/2 greater than 12.

13. Explanations will vary. Possibilities include:
    a.   7/9 + 6/13 is **greater than 1.** 7/9 is close to one and 6/13 is just less than a half.
    b.   4/6 + 6/7 is **greater than 1.** Both fractions are greater than 1/2.
    c.   4/5 – 3/4 is **less than 1/2** because the fractions are close to the same value.
    d.   3/7 + 2/9 is **between 1/2 and 1** because 3/7 is close to 1/2 adding 2/9 is less than 1/2.
    e.   4 1/5 – 3 1/20 is **greater than 1** because 4 – 3 = 1 and 1/5 is greater than 1/20.
    f.   7 3/7 – 6 2/3 is **between 1/2 and 1** because 7 – 6 is 1 but 3/7 and 2/3 are near 1/2 but 2/3 is greater than 3/7.

15. First we must find the sizes of all the pieces relative to the large triangles (a or b) represented by 1. Since parts a and b fit exactly on each other and **a = 1, b = 1** also. Now, c and e are identical, c and e fill d, c and e also fill g, and c, d, and e fill b. So **c = 1/4, e = 1/4, d = 1/2, and g = 1/2.** Since c, d, e, f, and g fill a + b, the sum 1/4 + 1/2 + 1/4 + f + 1/2 = 2. So **f = 1/2.**
    a.   So **b + c + d + e + f = 1 + 1/4 + 1/2 + 1/4 + 1/2 = 2 1/2.**
    b.   **a + d + e = 1 + 1/2 + 1/4 = 1 3/4.**
    c.   Since only a and g are missing, the figure has an area of 4, **4 – a – g = 4 – 1 – 1/2 = 2 1/2.**

17. Counterexample will vary. Possibilities include:
    a.   3/4 – 1/4 = 2/4 but 1/4 – 3/4 = –2/4.
    b.   (7/8 – 4/8) – 1/8 = 2/8 but 7/8 – (4/8 – 1/8) = 4/8.

19. a.   1/2 + 2 1/4 = 2 3/4; 1 1/2 + 1 1/4 = 2 3/4; 2 5/8 + 1/8 = 2 3/4
    b.   Think of a subtraction problem involving 2 3/4 and then reverse it to make an addition problem.

21. a.   The next two rows are: **1/6   1/30   1/60   1/60   1/30   1/6**
                                    1/7   1/42   1/105   1/140   1/105   1/42   1/7
    b.   Any 2 adjacent fractions have a sum equal to the fraction between them in the row above.

23. a.   Calculators perform computations in fraction form. 7/12 is the simplest form of the sum of 1/3 and 1/4.
    b.   Another press of '=' will display **19/12.**
    c.   Responses will vary. One solution is to count: 19/12 (5 times), 22/12 (6 times), ...34/12 (**10 times**).
    d.   **No.** There is no integer n such that $7 + 3n = 69$.

25. a.   1/5 – 2/9 = 9/45 – 10/45 = –1/45. **–1/45 is the result of the computation.**
    b.   Each time the '=' key is pressed another 2/9 = 10/45 is subtracted from the display. The next 2 presses will display **–31/45, then –41/45.**

c. Each time the '=' key is pressed the numerator decreases by 10. So counting down to −81 we have: −21 (3 presses), −31 (4 presses), −41 (5 presses), ... −81 (9 presses). NS will appear when the numerator and denominator have a common factor other than 1. This symbol indicates that the fraction in the display is not in simplest form.

27. Responses will vary. Possibilities include:
    a. Because 1/5 + 1/4 + 1/4 is about 3/4, the $720 is about 1/4 of the total. So the total was about 4 times $720 or about $2800.
    b.

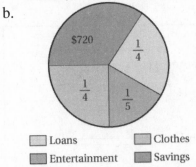

    Loans   Clothes
    Entertainment Savings

    c. Let *I* represent the inheritance. Then **[1 − (1/4 + 1/4 + 1/5)]*I* = 720.**
        (20/20 − 5/20 − 5/20 − 4/20)*I* = 720. So (6/20)*I* = (3/10)*I* = 720 and ***I* = 2400.**

29. Adding and subtracting integers and fractions is similar because the same properties and models apply. However, they are different because you often have to change a fraction to an equivalent form before you can work with it. The only time you have to do this with integers is when you regroup in order to subtract or add a result greater than 10. When you regroup with integers, there is a standard way to do it, but when you make equivalent fractions you have several choices.

31. **No.** Explanations will vary. One possibility is: The fractional raises are with respect to different wholes. If *I* represents the initial salary, the final salary, ***F***, is not *I* + (1/10)*I* + (1/20)*I* = *I* + (3/20)*I*. The final salary, ***F* is**; *I* + (1/10)*I* + 1/20[*I* + (1/10)*I*] = *I* + (1/10)*I* + (1/20)*I* + (1/200)*I* = ***I* + (31/200)*I*.**

33. The least sum from 1, 2, 3, 4 is **1/3 + 2/4 = 10/12 = 5/6.** The least sum from 5, 6, 7, and 8 is **5/7 + 6/8 = 82/56.** The least sum from 2, 5, 8, and 9 is **2/8 + 5/9 = 58/72 = 29/36.** With 0 < *a* < *b* < *c* < *d* the smallest sum is given by (*a*/*c*) + (*b*/*d*).

35. a. If we restrict the problem to the smallest positive difference from 1, 2, 3, and 4, we get **(2/4) − (1/3) = 1/6.** If we drop the restriction we get (2/3) − (4/1) = −3 1/3.
    b. The same strategy gives **(6/8) − (5/7) = 1/28** for 5, 6, 7, and 8.
    c. The smallest difference for 2, 5, 8, and 9 is **(5/9) − (2/8) = 11/36.**
    d. With 0 < *a* < *b* < c < d, the smallest positive difference is given by **(*b*/*d*) − (*a*/*c*).**

37. Responses will vary. To answer this question students will have to make assumptions about the dollar amounts spent subject to sales tax and the dollar amounts actually paid in income tax. Note that, except for allowable credits and deductions, all income is taxable as income but only some expenditures are subject to sales tax and that varies with respect to the types of purchases made. The dollars on which sales tax has been paid are also subject to income tax.

39. a.   Responses will vary. One possibility is: **7/12 = 1/3 + 1/4.**
    b.   Guess-check-revise leads to **2/7 = 1/4 + 1/28.**
    c.   Responses will vary. One possibility is: **3/5 = 1/2 + 1/10 = 1/3 + 1/5 + 1/15.**

SECTION 6.3

1.   The area of the square = 1.
    (a)

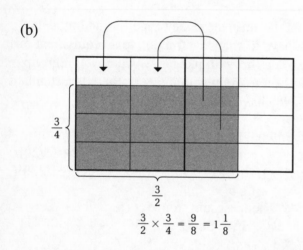

    (c)

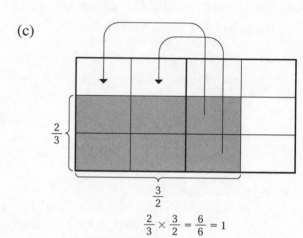

    (b)

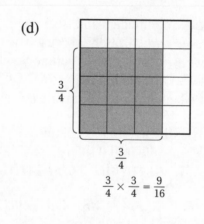

3.   a.   The shaded area represents **(3/5)** of **(2/5)**, a product of **(6/25)**.
    b.   The shaded area represents **(5/6)** of **(2/3)**, a product of **(10/18)**.

5.   Explanations may vary. A possibility is: **a and c may be represented by** $1/2 \times$ **10.** In parts a and c, a set of 10 is the whole which is to be divided into 2 equal subsets. In parts b and d the set of 10 is one of 2 equal subsets that, together, make up the whole.

7.   If 1/3 of your stock portfolio has decreased by 150%, the other holdings remaining constant, what is the change in the value of your holdings?

9.   The multiplicative inverse of a number x is the number y such that $(x)(y) = 1$.
    a. $(-7/9)(\mathbf{-9/7}) = 1$
    b.   2 5/7 = 19/7: $(19/7)(\mathbf{7/19}) = 1$

c. (1 1/2 + 3/4) 6 = 54/4: (54/4)(**4/54**) = 1.

d. $(-d/c)(-c/d) = 1$ unless $c = 0$ in which case there is no multiplicative inverse.

11. (a) $3\dfrac{1}{4} \div \dfrac{3}{4} = 4\dfrac{1}{3}$ (b) $1\dfrac{2}{6} \div \dfrac{2}{6} = 4$

13. a. I have 1/2 cup of sugar. The recipe calls for 2/3 cup. What fraction of a recipe can I make with the available sugar?

b. It usually takes me 1 1/2 hours to mow my lawn, but I only have an hour to do it right now. What fraction of my lawn will I be able to get mowed in an hour?

c. The stock market has dropped 3/4 of a point for many days in a row. Now it is 16 points below where it started. How many days has the marked dropped 3/4 point to get to this level?

15. The quotient by the complex fraction method is:
3/5 ÷ 2/3 = (3/5)/(2/3) = (3/5 × 3/2)/(2/3 × 3/2) = (9/10)/1 = 9/10.

17. The quotient by "invert and multiply" is: **3/5 ÷ 2/3 = 3/5 × 3/2 = 9/10.**

19. a. (5)(1/4) is **greater than 1** because (5)(1/5) = 1.

b. (7/5)(1/4) is **less than 1/2** because 7/5 is close to 1 and 1/4 < 1/2.

c. (1/2)(1/3) is **less than 1/2** because 1/3 <1.

d. (3/4)(1 1/3) is **between 1/2** and 1 because 3/4 < 1, but 3/4 > 1/2.

e. (1 1/2)(5/6) is **greater than 1** because 5/6 is about 1.

f. (9/10)(4/5) is **between 1/2** and 1 because both fractions are just a little less than 1.

21. The Fundamental Law states that for $c \neq 0$, $(a/b)(c/c) = ac/bc$. The multiplicative identity for a rational number a/b is b/a. Both these concepts involve multiplying a rational number by another rational number. In one instance the multiplication is to produce 1, in the other instance the multiplication is by 1.

23. a. $(4/3)x = 12$. $(3/4)(4/3)x = x = (3/4)12 = $ **9**.

b. $8y = 27$. $(8)(1/8)y = (1/8)27 = y = $ **27/8 = 3 3/8**.

c. $12/x = 4/5$. $x/12 \times 12/x = 1 = x/12 \times 4/5 = 4x/60 = x/15$. $x = 15$.

d. $-r/5 = 16/3$. $(-5) -r/5 = 16/3x -5 = r = -80/3 = -26\ 2/3$.

25. a. Another press of the '=' sign will display **256/625**.

b. The display would show the sequence 4/5, 16/25, 64/125, 256/625, 1024/3125, 4096/15,625 for **6** presses of the 'equal' key.

c. The display is 0.262144. The calculator is incapable of displaying fractions with denominators that length. Therefore it switches to decimal display.

d. The display would be **0** because 0 × 4/5 = 0.

27. a. 3 2/9 ÷ 5 1/7 is between 1/2 and 1 because the dividend is more than half of the divisor.

b. 14 ÷ 2/3 is greater than 1 because there will be more than 1 group of 2/3 in 14.

c. 56/10 ÷ 49/10 is greater than 1 because there will be more than 1 group of 49/10 in 56/10.

   d. 3/4 ÷ 2/3 is greater than 1 because 2/3 is slightly smaller than 3/4.

   e. 1 ÷ 8/3 is less than 1/2 because 8/3 is greater than 2.

   f. 1/10 ÷ 1/5 is 1/2 because 1/5 is half of 1/10.

29. Responses will vary. Examples are:

   a. (6)(10)(1/6) = 10; (5)(11)(10/55) = 10; To find the factors pick any 2 non-zero numbers, say a and b, as the first 2 factors. To find the third factor, f, solve the equation $abf = 10$.

   b. The 3 factors could be length, width, and height of a right-angled box with a volume of 10.

31. a. Every time the '=' key is pressed the display is divided 3/4. So the next press of '=' will display the result of 64/27 ÷ 3/4 = **256/81**.

   b. The sequence of displays is 4/3, 16/9, 64/27, 256/81, 1024/243, 4096/729. So 4096/729 appears on the **sixth** press of '='.

   c. The result would be 0 because $0 ÷ n = 0$.

33. Responses will vary. Possibilities include:

   a. Let R represent the ribbon remaining. Then R = 25 1/3 – [(1/3)(25 1/3) + (1/2)(1/3)(25 1/3)]. This equation represents the beginning total length reduced by the total amount of ribbon sold.

   b. As an alternative approach, begin with the amount of ribbon given to each store, 1/3(25 1/3)]. Let L represent the amount left after the various sales. The amount remaining is all the ribbon given one store, 1/3(25 1/3), and half the ribbon given another, 1/2[(1/3)(25 1/3)]. So L = 1/3(25 1/3) + 1/2(1/3)(25 1/3).

   c. Solving the equation in b, L = 1/3(76/3) + 1/6(76/3) = 76/3(1/3+1/6) = 76/3(3/6) = 76/6 = **12 2/3 yd left.**

   d. 25÷3 ≈ 8, about the amount of ribbon given each store. Since one store sold none, its 8 yards were left. Another store sold half of its 8 yards, leaving 4. Because the third store sold all its ribbon, the amount left is about 12 yards.

35. Responses will vary. Possibilities include:

   a. The conjecture that 'multiplication makes numbers bigger' most probably arises from experience with natural numbers. For example the product of 3 and 4 is larger than either 3 or 4.

   b. The conjecture is true if multiplication is restricted to the set of natural numbers.

   c. It is false when multiplying whole numbers ($0 \times 6 = 0$), integers ($-2 \times 3 = -6$) and rational numbers ($1/2 \times 1/3 = 1/6$).

37. Responses will vary. A possibility is:

(2 3/5) × (1 1/4) = (2 + 3/5)(1 + 1/4)	convention for the representation of mixed numbers.
= 2(1) + 2(1/4) + 3/5(1) + (3/5)(1/4)	distributive property
= 2 + 1/2 + 3/5 + 3/20	rational number multiplication
= 2 + 10/20 + 12/20 + 3/20	Fundamental Theorem of Fractions
= 2 + 25/20 = **3 1/4**	arithmetic simplification

39. a. The least product is (1/3)(2/4) = (1/4)(2/3) = **2/12 = 1/6**.

   b. The least product is (6/7)(5/8) = (5/7)(6/8) = **30/56 = 15/28**.

   c. The least product is (2/8)(5/9) = (5/8)(2/9) = **10/72 = 5/36**.

   d. If $0 < a < b < c < d$, the least product is given by **(ab)/(cd)**.

41. Responses will vary. Possibilities include:
    a. If we are restricted to the set of natural numbers, then division of a number generally results in a smaller number. The exception is division by 1. Thus the conjecture that division, generally, makes numbers smaller.
    b. The conjecture is true within the integers in all cases in which the dividend is positive and the divisor is greater than 1 (example: $6 \div 3 = 2$) or when the dividend is positive and the divisor negative (example: $6 \div -2 = -12; -12 < 6$).
    c. It is false within the integers if both dividend and divisor are less than 0, for example $-6 \div -2 = 3$ and $3 > -6$. It is also false when the dividend is less than zero and the divisor is greater than 1, for example $-6 \div 2 = -3: -3 > -6$.

43. a. i. $1 \div 3/4 = \textbf{4/3}$ ii. $1 \div 5/6 = \textbf{6/5}$ iii. $1 \div 3/7 = \textbf{7/3}$ iv. $1 \div 4/9 = \textbf{9/4}$.
    b. The pattern suggests that $\textbf{1} \div \textbf{a/b} = \textbf{b/a.}$
    c. $1 \div a/b = 1/(a/b) = [1/(a/b)] \times [(b/a)/(b/a)] = [1(b/a)]/[(a/b)(b/a)] = (b/a)/[(ba/ab)] = (b/a)/1 = b/a$ for all non-zero integers a and b.

45. a. The least quotient is $(2/3) \div (4/1) = \textbf{1/6}$.
    b. The least quotient is $(6/7) \div (8/5) = \textbf{15/28}$.
    c. The least quotient is $(5/8) \div (9/2) = \textbf{10/72} = \textbf{5/36}$.
    d. If $0 < a < b < c < d$ the least quotient is $(\textbf{b/c}) \div (\textbf{d/a}) = ba/dc$ or the equivalent $(\textbf{a/d}) \div (\textbf{c/b})$.

47. In one school the ratio of participants to school population is 156/723, approximately 150/750 or 1/5. In the other the ratio is 285 to 1208, about 300 to 1200 or 1/4. So the larger school had the better participation. Conversion to decimals, $156/723 \approx .22$ and $285/1208 \approx .24$ confirms the estimates.

49. (a) $\dfrac{1}{6} \times \dfrac{1}{6} = \dfrac{1}{36}$ (b) $\dfrac{1}{50} \times \dfrac{1}{49} = \dfrac{1}{2450}$

    (c) $0.1 \times 0.1 \times 0.1 \times 0.1 = 0.0001$

SECTION 6.4

1. (a)

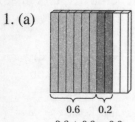

0.6   0.2

0.6 + 0.2 = 0.8

(b)

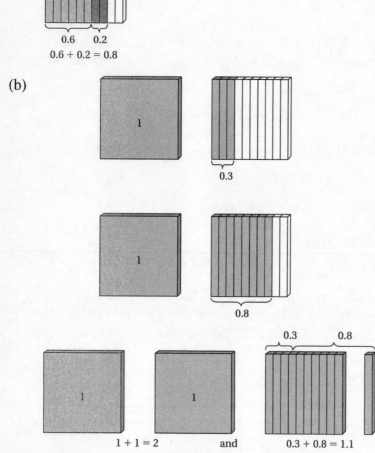

0.3

1

0.8

1 + 1 = 2   and   0.3 + 0.8 = 1.1

1.3 + 1.8 = 3.1

(c)

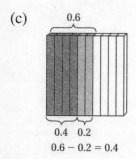

0.6

0.4   0.2

0.6 − 0.2 = 0.4

(d)
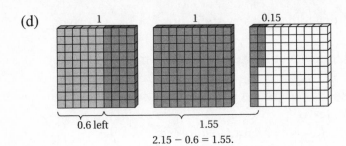

0.6 left   1.55

$2.15 - 0.6 = 1.55.$

3. The area of the square = 1.

(a)
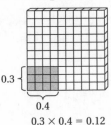

0.3

0.4

$0.3 \times 0.4 = 0.12$

(b)
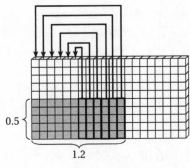

0.5

1.2

$1.2 \times 0.5 = 0.6$

5. a *i*. $24.6 + 3.09 = 24 + 6/10 + 3 + 9/100 = 240/10 + 6/10 + 300/100 + 9/100 = 2400/100 + 60/100 + 300/100 + 9/100 = (2400 + 60 + 300 + 9)/100 = 2769/100 = \textbf{27.69}$

a *ii*. **24.60**
  +3.09
  27.69

b. Because the same 2 numbers were added in both cases and because the sum is a unique number, **the sums are the same**. Discussions will vary but students may note that the decimal algorithm seems quicker and less prone to error.

7. a *i*. $24.6 \times 3.09 = (24 + 6/10) \times (3 + 9/100) = (240/10 + 6/10) \times (300/100 + 9/100) = (240 + 6)/10 \times (300 + 9)/100 = (246/10) \times (309/100) = 76014/1000 = 76.014$

a *ii*. **24.6**
  ×3.09
   22 14
  000 0
  738
  76.014

b. The products are the same because the same two numbers were multiplied in both cases and because the product is a unique number. Discussions will vary but students may note that the decimal algorithm seems quicker and less prone to error.

9. a. If I drive 1.23 miles to work and then 0.8 miles to the grocery store, how far have I driven?
        1.23 + 0.8 = 2.03 miles
    b. I have $1.05 in my wallet, and I find $0.20 in my coat pocket. How much money do I have?
        $1.05 + $0.20 = $1.25
    c. Julio ran the race in 1.23 minutes and Janika ran the race in 0.8 minutes. By how many minutes
        did Janika beat Julio? 1.23 − 0.8 = 0.43 minutes
    d. I had $1.05, but I bought a sticker for $0.20. How much money do I have left?
        $1.05 − $0.20 = $0.85

11. a. It is close to 1.5 because using rounding, we get 1.5 + 0.5 − 0.5 = 1.5.
    b. It is close to 0 because 3.7 and 4 are close to each other.
    c. It is close to 3 because 0.4 + 1.5 + .125 is close to 2. Adding 0.7 more gets us close to 3.
    d. It is close to 2.5 because the first term and last term are both close to 11, so adding 11 and
        subtracting 11 is zero. Thus, using rounding we have 12.5 − 10, which is 2.5.

13. Responses will vary. A possibility is: By lining up the decimal points the numbers in a column are
    all multiples of the same power of ten. Thus by adding up the numbers in the columns the
    distributive law is being applied.

15. a. 75.02 + 0.01; 74.01 + 1.02; 70.01 + 5.02
    b. Select one addend and then subtract from 75.03 to find the other addend.

17. Responses will vary. Possibilities include:
    a. (31.8)(42) = **1335.6** because (30)(40) = 1200.
    b. Because 0.3 of 4 is just over 1, (0.318)(4.2) = **1.3356**.
    c. (318)(0.42) = **133.56** because (300)(0.5) = 150.
    d. Because (3)(4) = 12 the product of 3.18 and 4.2 is **13.356**.

19. Reasoning will vary. Possibilities include:
    a. 691.04 ÷ 5.6 = **123.400** because 500 ÷ 5 = 100.
    b. 69.104 ÷ 56 = **1.23400** because 60 is bigger than 50.
    c. 0.69104 ÷ 0.56 = **1.23400** because .6 is greater than .5 but not double.
    d. 691.04 ÷ 56 = **12.3400** because 600 ÷ 50 is greater than 10.

21. a. Let S represent the standard cost and let m represent time on the phone in minutes. The equation
        used is $S = 24.95 \div 0.35\,m$.
    b. Let E represent the economy cost. Then $E = 30.00 + (m - 30)(0.38)$.
    c. The chart displays costs under the two plans as a function of time.
    d. Responses will vary. The spreadsheet does not provide a solid basis for choice. However, the
        economy plan costs are based on peak time usage and can generally be expected to be lower than
        the values in the table. If the phone is to be used for calls other than emergencies, the economy
        plan is better.

23. a. The number of square feet may be determined by dividing the asking price by the price per square foot.

   b. The square footages are, respectively, **1532, 1198, 1689, 1280, 1598, 3190**.

   c. **E3 = B3 ÷ D3, E4 = B4 ÷ D4, ... , E8 = B8 ÷ D8**

   d. i. The costs in dollars per square foot are now, respectively, **52.81, 53.84, 42.92, 57.03, 53.75, 55.76**.

      ii. **F3 = (B3 − 2000) ÷ E3, F4 = (B4 − 2000) ÷ E4, ... , F8 = (B8 − 2000) ÷ E8**

   e. **B9 = (B3 + B4 + B5 + B6 + B7 + B8)/6**. When one of a data set is much higher or lower than the rest of the data an arithmetic average is not appropriate because that one value has a disproportionate effect on the average. These conditions apply with this data.

25. a. $2.3 ÷ 5/6 = 4/5$. To find a solution in decimal form, we must truncate the repeating decimals to, for example, $0.66 ÷ 0.83 = 0.795$. The fraction computation yields a more accurate answer because the answer with decimals is just an approximation.

   b. $1/2 ÷ 3/5 = 5/6$; $0.5 ÷ 0.6 = 0.0.83333...$ Both computations yield the same result.

27. Answers will vary.

SECTION 6.5

1. a.

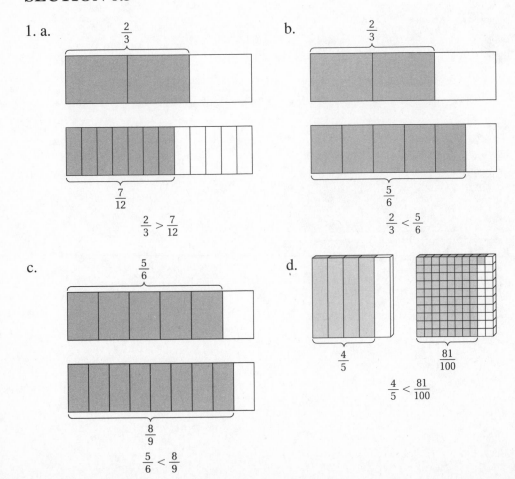

3. (a) Let a flat = 1.

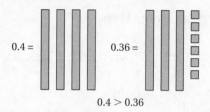

0.4 =   0.36 =

0.4 > 0.36

(b) Let a large cube = 1.

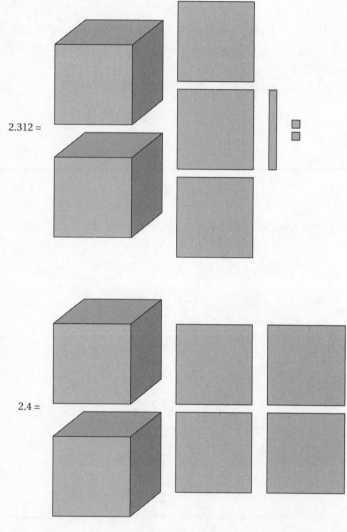

2.312 =

2.4 =

2.312 < 2.4

5. a. $8/25 = 0.3\underline{2} > 0.3\underline{1}$
    b. $3/4 = 0.\underline{75} < 0.\underline{8} = 4/5$
    c. $-\underline{1}.85 > -\underline{2}$
    d. $-0.23\underline{23}\ldots > 0.23\underline{00}\ldots$

7. a. 3/4 = 36/48 < 37/48 < 38/48 < 39/48< 40/48 = 5/6
    b. −1.4 < −1.39 < −1.38 < −1.37 < −1.3
    c. 5/7 = 50/70 < 51/70 < 52/70 < 53/70 < 60/70 = 6/7
    d. −3/5 < −1/5 < 0 < 1/5 < 3/5

9. Responses will vary. Some possibilities are:
    a. 0.15, 0.222... , 0.25   b. 1/8, 0.125125... , 3/8
    c. −1/3, 0.111... , 1/3   d. −2.02, −2.01919..., −2.01

11.   Answer for Exercise 11

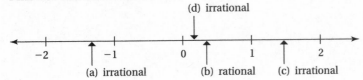

13. a. 42,5000
    b. 0.000276
    c. 0.00000103
    d. 8,220

15. Responses will vary. A possibility is: Suppose a/b is a rational number with a reciprocal between 0 and 1. Then $0 < b/a < 1$. b/a is less than 1, the reciprocal of b/a, $1/(b/a) = a/b$, is greater than 1. Thus any number greater than 1 has a reciprocal between 0 and 1. So the set of numbers with reciprocals between 0 and 1 is all the rational numbers that are greater than 1.

17. 1/3 < 1/2 because 1/3 + 1/6 = 1/2

19. Yes, the sum of two irrational numbers is always an irrational number because the set of rational numbers and the set of irrational numbers are disjoint sets (see Figure 6.25). Thus, adding two numbers in the set of irrational numbers cannot result in a number in the set of rational numbers.

21. a. Let $n = 0.11...$ and $10n = 1.111...$ Then $9n = 1$ and $n = 1/9$. So **0.22... = 2(1/9) = 2/9, 0.33.. = 3/9, 0.44.. = 4/9, 0.55... = 5/9 It appears that 0.aa... = a/9**.
    b. Let $n = 0.12...$ and $100n = 12.12...$ So $99n = 12$ and $n = 12/99 = .12$. By similar computations $0.23... = 23/99$, and $0.34... = 34/99$. It appears that for **b = a + 1, 0.ab... = ab/99**.

23. Responses will vary. A possible argument is: *Right next to* implies that there is nothing in between. But 0 is a rational number and the denseness property of rationals states that between any 2 rationals there is another rational number. So suppose that a/b is right next to 0. But there is another number between a/b and 0. So a/b is not right next to 0. So then imagine that c/d is right next to 0. But there is another rational between c/d and 0. Thus there is no number right next to zero.

25. The sequence is **increasing**. Explanations will vary. One argument is: The first terms of the sequence are 0, 1/2, 2/3, 3/4, 4/5. These are increasing. Now, as n increases the numerator becomes closer in value to the denominator and the value of the fraction becomes closer and closer to 1.

27. Responses will vary. A possibility is: We must look at three cases: the first, $a/b < 0$; the second $0 < a/b < 1$; the third $a/b > 1$. In the first case, because the square of a number is always positive the square of any number < 0 is greater than the number. Now, for the second case: if $a/b < 1$ then $(a/b)^2 = (a/b)(a/b)$. Since $a/b < 1$ then (a/b of a/b) is only a part of a/b and thus is less than a/b. In the third case, if a/b is > 1 then a/b can be written as a mixed number, say $c + d/e$. Now, $(c + d/e)(c + d/e) = c^2 + 2c(d/e) + (d/e)^2 = c^2 + c(d/e) + c(d/e) + (d/e)^2 = c(c + d/e) + (d/e)^2 > c + (d/e)$. So the square of a positive mixed number is greater than the number.

29. Answers will vary but might include height, times in track events, gas prices, grade point averages, test scores, or representations of large numbers in the newspaper (e.g., 1.8 billion).

CHAPTER 6 REVIEW EXERCISES

1. a.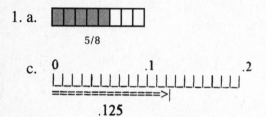
   5/8

   b.
   2 3/5

   c. 0　　.1　　.2
   .125

3.

Fraction form	Decimal form	Method of Solution
5/12	**0.4166...**	by calculator
125/999	0.125...	$n = 0.125...$, $1000n = 125.125...$, $999n = 125$, $n = 125/999$
56/100	0.56	definition of decimal notation

5.

	Number	Additive Inverse	Multiplicative Inverse
a.	3/4	–3/4	**4/3**
b.	5 1/2	–5 1/2	**2/11**
c.	**–3.2**	**3.2**	**1/–3.2 = –0.3125**
d.	0	**0**	**not defined**

7. a. $1 \frac{1}{3} \times -5 \frac{1}{2} = 4/3 \times -11/2 = \mathbf{-44/6} = \mathbf{-7\ 1/3}$
   b. $4 \frac{1}{5} \times 2 \frac{1}{3} = 21/5 \times 7/3 = 147/15 = \mathbf{9\ 4/5}$
   c. $0.7 \times 3.16 = \mathbf{-2.212}$
   d. $4.08 \times 3.5 = \mathbf{14.26}$

9. Responses will vary. Some example word problems are:
   a. One brand of instant potatoes calls for 1 3/4 cup of water and 1 1/2 cup milk. How much liquid does the recipe require? S = 1 3/4 + 1 1/2 = **3 1/4**.

   1 3/4　+　1 1/2　=　3 1/4

   1 3/4　-　1 1/2　=　1/4

b. In the above recipe how much more water than milk is required? **d** = 1 3/4 – 1 1/2 = **1/4.**

c. A rectangular piece of wood is 1 3/4 yd 1 1/2 yd. What is the area? **p** = 1 3/4 × 1 1/2 = **2 5/8 sq yd**.

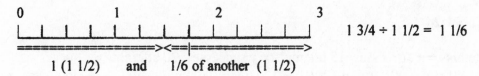

1 3/4 × 1 1/2 = 1 + 3/4 + 1/2 + 3/8 = 2 5/8

d. It has been determined that 1 1/2 revolutions of a crank on a car jack can lift a car 1 3/4 in. How many revolutions are required to lift the car 1 in? **q** = 1 3/4 ÷ 1 1/2 = 7/4 × 2/3 = **14/12 = 1 1/6**

```
0              1          2          3
|_|_|_|_|_|_|_|_|_|_|_|_|_|          1 3/4 ÷ 1 1/2 =  1 1/6
=================><<==|==============>
   1 (1 1/2)    and   1/6 of another (1 1/2)
```

11. 2.357... **is a rational number.** Every rational number can also be represented as either a terminating decimal or, as in this case, a repeating decimal.

13. a. Methods of estimation will vary. One is: the sum of the weights
S = (1/2 + 6 3/4) + 2 3/4 + 2 1/8 + 1 is approximately 7 + 3 + 2 + 1 = 13. So the groceries should be put into 2 sacks.

b. Responses will vary. One possibility is: Since the total is about 13 lbs and the ham weighs about half of the total, put the ham in one bag and the remainder of the groceries in another bag.

15. a. Explanations will vary. One possibility is: Martha's statement is **false.** If her old salary is represented by S and her new salary by N then $N = (3/4) S$. Expressing her new salary as a difference: $S – (1/4) S = (3/4)S$. So her salary was **reduced by 1/4.**

b. Since her new salary, N, is $(3/4) S$ we have $N = 3/4 S$ or $S = (4/3)N = N + (1/3)N$. So her new salary would have to be increased by a third to again be equal to her original salary.

17. Responses will vary. A possibility is; **Both are correct**. When a division by 5 produces the result 32 R 2 it means 32 groups of 5 and 2 ungrouped units. A display of 32.4 when the quotient of a division by 5 means 32 groups of 5 and 4/10 of a group of 5. But 4/10 of a group of 5 is (4/10)5 = 2. So the displays are different representations of the same quotient.

19. Responses will vary. If one number is negative and the other positive then the positive number is the larger. If the numbers are of the same sign an efficient process is to use a calculator to represent both rationals as decimals. Compare the decimals place by place until there is a difference. If both are positive then the larger of the place values identifies the larger number. If both are negative the larger of the place values identifies the smaller number. If the calculator does not show a difference represent both as rationals with a common denominator and compare the numerators as the place values were described.

21. Responses will vary depending on student-student interactions.

SECTION 7.1

1. The ratio girls: boys = **2:3**. The ratio students: boys = **5:3**; the ratio students: girls = **5:2**.

3. 15 cm to 5 cm = 15:5 or **3:1**.

5. 4 in to 12 in = 4:12 or **1:3**.

7. 4 kg to 10 kg = 4:10 or **2:5**.

9. 6 into 12 ft = $\frac{1}{2}$:12 = **1:24**.

11. 8 successes: 10 attempts = x successes: 15 attempts = y successes: 20 attempts = z successes: 75 attempts. So 8/10 = x/15; x = (8/10)15 = 120/10 = **12 successes**. Similarly, y = **16 successes**, z = **60 successes**.

13.

c	d	
6	2	c/d = 3/1 = 6/d. So 3d = 6 and d = 6/3 = 2.
9	3	c/d = 3/1 = c/3. So c = 3 × 3 = 9.
10	**10/3**	c/d = 3/1 = 10/d. So 3d = 10 and d = 10/3.
27	9	c/d = 3/1 = c/9. So c = 9 × 3 = 27.
21	**7**	c/d = 3/1 = 21/d. So 3d = 21 and d = 21/3 = 7.
12	4	c/d = 3/1 = c/4. So c = 12.

15.

x	1.8	3	**6**	**9**	4.5	**15**
y	**3**	**5**	10	15	**7.5**	25

17. Let R represent the ratio 180 successes to 210 attempts. Then R = 180 ÷ 210 = **0.857** by calculator.

19. The ingredients flour (F), milk (M), and baking powder (P) are in the ratio: F:M:P = 8:1 1/3:5. Letting F = 6 we have 6:M:P = 8:1 1/3:5 or individually 6:M = 8:1 1/3 and 6:P = 8:5. Solving the first of these proportions we have: 6:M = 8:1 1/3; 6(1 1/3) = 8 = 8 M. So we need **1 cup of milk**. Solving the second we have: 6:P = 8:5; 8P = 30; P = 30/8 = 3 3/4. So we also need **3 3/4 tablespoons of baking powder**.

21. 77 miles

23. More than 10 miles since the map covers a larger region.

25. 0:30 = $\frac{60}{30}$ = $\frac{2}{1}$.

27. 40:240 = 1:6 = $\frac{1}{6}$: 1

29. With a win/loss ratio of 31 to 26 San Francisco had a winning ratio of 31/(31 + 26) = **31/57**.

31. The unit price is $1.50/ 3 or **$0.50 per grapefruit**.

33. The unit price is $1.60/.5 kg = **$3.20 per kg**.

35. If there are 32 students and 8 are male, then 24 are female. So the ratio of males to females is 8:24, which is 1:3 in simplest form.

37. The ratio by weight is 5 parts blue grass per 3 parts clover seed. So 5/8 of the total weight of a lot of seed should be blue grass and 3/8 clover. Thus an 8 lb bag would have 5 lbs blue grass and 3 lbs clover. Maintaining this ratio, a 16 lb bag would have twice as much of each because 16 is twice 8, a 24 lb bag 3 times as much of each and a 40 lb bag 5 times as much. Thus:

lbs. total	8	16	24	40
Blue grass (lbs)	5	10	15	25
Clover seed (lbs)	3	6	9	15

39. Alisha's mileage is 250 miles per 12 gallons or 20.83...miles per gallon. Nicole's mileage is 300 miles per 15 gallons or 20 miles per gallon. **Alisha** gets slightly better mileage.

41. Arguments will vary from conjectures obtained from substituting several sets of numbers to simulate the current average to the following general argument: The current average, C, is 0.305 which is the ratio of hits, h, to at bats, a. The predicted average, P, is the ratio of $h + 7$ to $a + 20$. We want to compare P to C. Suppose that $P = xC$. If x is greater than 1, then $P > C$. So $(h + 7)/(a + 20) = 0.305x$. Since $h/a = .305$, $h = .305a$ and $(.305a + 7)/(a + 20) = 0.305x$. So $x = (.305a + 7)/(.305a + 6.1)$. Since this last expression is greater than 1 for all a, x is greater than 1 and P is greater than C. So **Marge is correct**.

43. Four bars of soap would be a better buy for any price, P, that would make the ratio $P/4$ less than the ratio 1.70/6. Both ratios are unit prices and the lower unit price is the better buy. So: $P/4 < 1.70/6$ if $P/4 < 0.283...$ or if $P < 4.(283..)$. So if the price of 4 bars is **$1.13 or less**, the four bars are a better buy.

45. Since the ratio of girls to boys in Mr. Maddox's class is 3:2 the ratio of girls to students is *3/5*. Thus there are 18 girls in this class. The ratio of girls to students in Ms. Wood's class is 2 to 3. Thus of her 27 students 18 are girls. So **there are the same number of girls in both classes.**

47. The ratio $(1/3):(1/2) = (1/3) \div (1/2) = (1/3) \times (2/1) = $ **2:3**.

49. The bars that are given represent 2.5(100) miles per 10 gallons and 5(100) miles per 20 gallons: 25 miles per gallon. So, letting x and y represent the heights of the missing bars, $100x/30 = 25$ and $100y/40 = 25$. So bar x should be 7.5 and bar y should be 10.

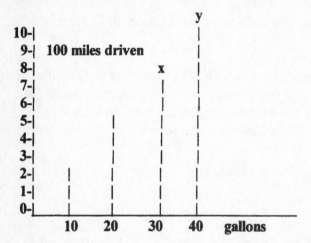

51. **Jackie's results are acceptable.** The ratio of RCA to Sony to Panasonic 3:5:4. It is possible that one person owns more than 1 television and that these are different brands. There could easily be 12 sets owned among 10 persons. It might clarify matters if Jackie referred to the ratio of RCA sets *owned* to Sony sets *owned* rather than to the ratio of RCA set owners and Sony set owners.

53. Responses will vary. The facts are that the ratio of firearms per owner is 4.5:1. If there are 50 million firearm owners then there are about 225 million privately owned firearms in the country.

55. Specific responses will vary. The facts are: The sales values, in thousands of dollars, are about 15 per 1 year, 30 per 2 years, 45 per 3 years and 60 per 4 years. The ratio of sales to time in years is constant. This graph shows no growth (or decline) in sales.

SECTION 7.2

1. If 2 quantities, say y and x, vary proportionally then $y = kx$, k a constant, for all values of y and x. So, since $3 = 3(1)$, $k = 3$ and we see that for all values of x, $y = x$. So, completing the table we have:

x	1	1.5	2	2.5	3	3.5
y	3	4.5	6	7.5	9	10.5

3. If $2/5 = 4/10$, then $10/4 = 5/2$ by the **reciprocal property**.

5. If $c/b = m/x$, then $(b)(m) = (c)(x)$ by the **cross product property**.

7. Estimates will vary. A possibility is: Because 60 is half again as big as 40, x should be half again as big as 9. So x is 9 plus half of nine, about 13. Solving for x, $x = (60/40)(9) = $ **13.5**.

9. a. By the cross product property we have $15x = 20$, $x = 20/15 = $ **4/3**. Alternatively; $(4/15)(5/1) = (x/5)(5/1)$; so $x = 20/15 = 4/3$. There is little to choose between in the two methods.

   b.   Again, by cross products we have $25x = 28$, $x = \mathbf{28/25}$ and $(x/7)(7/1) = x = (4/25)(7/1) = 28/25$.

   c.   By cross products we have $9(2x + 5) = 7(27)$. $18x + 45 = 189$ and $\mathbf{x = 8}$. Alternatively, $(9/7) = 27/(2x + 5)$; $(9)(3)/(7)(3) = 27/21 = 27/(2x + 5)$. And $21 = 2x + 5$ and $x = 8$.

11. $x = 72$

13. $x = 28$

15. $x = 9\dfrac{1}{3}$

17. $x = 35$

19. $x = 3$

21. $x = \dfrac{11}{8}$

23. $x = \dfrac{14}{9}$

25. $x = 2.5$

27. $x = \dfrac{48}{15}$ or $\dfrac{16}{5}$

29. $x = 14$

31. The lengths of corresponding sides of these rectangles do not vary proportionally because the length to length ratio is not equal to the corresponding width to width ratio. 10 to 5 is not equal to 20 to 15.

33. Answers will vary.

35. $AC = 6''$; $DF = \dfrac{14}{3}''$

37. a.   **No.** The x and y values do not give a constant k for y divided by x.

    b.   **Yes.** Every y value is -1 times the x value so $y = kx$.

    c.   **No.** The ratios of the various x, y pairs are not equal.

39. If the 40 spaces for full-sized cars were to be converted to compact spaces each space would be equivalent to 1 1/2 compact spaces. So the 40 full-sized spaces would become $40(1\ 1/2) =$ **60 compact spaces**.

41. Assuming proportionality: $2.70/6 bars = cost/10 bars; cost is **$4.50.**

43. We note that 15,000 is 15 times 1000 and, assuming proportionality between cost and distance, it will cost 15 times as much to drive 15,000 miles as 1000 miles. So the cost for 15,000 miles is $15(42) = $**630.**

45. Assuming proportionality, we have 7 ads/3 pages = ads/150 pages. The number of ads per issue is **350**.

47. Responses will vary. Possibilities include: Assume that the number of tickets, N, is proportional to the cost, C. Then $N/C = 3/5 = N/55$. So $N = (3/5)55 =$ **33 tickets**. An alternative approach is to determine the unit price, $5 / 3$ tickets = $1 2/3 per ticket. Now divide the unit price into the available money: $55 ÷ $1 2/3 per ticket = 33 tickets.

49. The U.S. dollars traded and the number of Euros received do **not** vary proportionally. Compare the numbers in the first and last columns in the table below.

US dollars traded	No. of Euros converted	Service charge	No. of Euros received
50	42.5	2.0	40.5
100	85.0	2.0	83.0
200	170.02	2.0	168.0
300	255.0	2.0	253.0
400	340.0	2.0	338.0
500	425.0	2.0	423.0

51. Arguments will vary. A possibility is: As a preliminary test of the claim substitute some numbers: let $a = 2$, $b = 5$, $c = 4$, and $d = 10$. So $2/5 = 4/10$ and $2/(2 − 5)$ does equal $4/(4 − 10)$ because both are equal to $− 2/3$. This is not a proof but it does support the claim. Now, suppose that $a/b = c/d$. Then $b/a = d/c$ by the reciprocal property. Continuing: $1 − (b/a) = 1 − (d/c)$ and $(a/a) − (b/a) = (c/c) − (d/c)$. So $(a − b)/a = (c − d)/c$. Again applying the reciprocal property we have $a/(a − b) = c/(c − d)$. Thus **Kay is correct**. (Note: This is correct only if $a ≠ b$, $c ≠ d$, and a, b, c and $d ≠ 0$, otherwise there is division by 0.)

53. Let G represent units of gasoline and let L represent units of oil. The mix $G:L$ is to be in the ratio 50:1. To figure out how many ounces of oil to use with 2 gallons or 256 ounces of gas, solve $256/L = 50/1$. $L = 5.12$ ounces of oil.

55. Responses will vary. A possibility is: The engineer applied the reciprocal property to get $t/12 = 50/210$. Then she solved for t by multiplying both sides by 12 obtaining $t = (50/210)12 = 2.857....$ The solution **is correct**.

57. Responses will vary. One possibility is: If 3 inches represents 20 miles then 12 inches, 4 times 3 inches, represents 4 times 20 miles or 80 miles and 18 inches, 6 times 3 inches, represents 120 miles. Thus the photograph represents an area 80 miles by 120 miles or 80 times 120 = **9600 square** miles.

59. Responses will vary. One possibility is: The ratio of glycerin to water is 2 to 3. This means that the ratio glycerin to solution is 2 to 5 or the solution is 2/5 glycerin and 3/5 water. Two fifths of 30 ounces is 12 ounces. The rest of the solution, $30 − 12$ ounces = 18 ounces, is water. To make the solution mix **18 ounces of water with 12 ounces of glycerin.**

61. Responses will vary. Possible arguments include: Since the ratio 4/4 is equal to the ratio 10/10, the frames are similar in shape and the picture will not have to be cropped. However, since 4 by 4 is not similar to 8×10 (the length ratio is different from the width ratio) some of the picture will have to be cropped.

63. Responses will vary. A possible argument is:

Squares	1	2	3	4	5
Corners	4	7	10	13	16

The number of corners **is not proportional** to the number of squares because the ratio corners to squares is not constant.

65. Responses will vary. One argument is: To each side of the proportion add 1 in the form of b/b and d/d. Thus since $a/b = c/d$, $(a/b) + (b/b) = (c/d) + (d/d)$. So $(a + b)/b = (c + d)/d$.

67. Responses will vary. Possibilities include:
    a. **Correct** because to determine amount of sales tax, T, the price, P, is multiplied by a constant, the tax rate, r. So $T = r P$, one of the definitions of proportional variation.
    b. **Generally this is incorrect.** Although potentially correct for the first several months of growth, one would not have to collect data for too long a time to find that the ratio of height to weight is not a constant.
    c. Although there is a rule of thumb that weight gain is proportional to excess calories, weight/calories = 1 lb/3500 cal, it is only a rule of thumb. Other variables, for example, exercise, affect the gain of weight.
    d. For a fixed unit price, u, this **is correct.** Value $= u$ times weight.
    e. This is **correct.** Because distance around, C, is equal to π times diameter, d, (distance across the circle) the ratio C to d is a constant. This is a definition of proportional variation.

69. Responses will vary. One possibility is: Simplifying after each step: cross multiply. Subtract the smaller value of x from both sides. Add the opposite of the number on the side with the x to both sides. Finally divide both sides by the number multiplying the x. So beginning with $(x + 3)/8$ $= (2x - 4)/4$ cross multiply and simplify: $4(x + 3) = 8(2x - 4)$, $4x + 12 = 16x - 32$. Since $4x < 16x$ subtract $4x$ from both sides: $4x + 12 - 4x = 16x - 32 - 4x$ and simplifying we now have: $12 = 12x - 32$. Adding the opposite of the number on the 'x side' we have: $12 + 32 = 12x - 32 + 32$, $44 = 12x$. Finally dividing by 12 we get $x = 44/12 = 11/3$. As a check we substitute 11/3 for x in both sides of the original equation: $(11/3 + 3)/8 = (20/3)/8 = 20/24 = 5/6$ and $[2(11/3) - 4]/4$ $= (22/3 - 12/3)/4 = 10/12 = 5/6$.

71. Responses will vary. Possibilities include: In 1972 and 1991 respectively we have the proportions 10.9/1000 = total marriages '72/1,000,000 and 9.4/1000 = total marriages '91/1,000,000. So the totals are: for 1972: 10, 900 and for 1991: 9,400. The difference is **−1,500 marriages.** The data reflects averages over a population and there very well may be wide variations in marriages over different population centers.

73. Solutions will vary. Possibilities include:
    a. 50/80 =10/x = 5/8; 5x = 80, x = 16. So **16 men** can build the 80 rods of wall.

b. Since 10 men can build 50 rods of wall in 12 days, then 1 man can build 5 rods of wall in 12 days and 1 man can build 5/12 of a rod of wall in 1 day. Since we want to build 80 rods of wall, if it were to be done by one man it would take 80/(5/12) days = 960/5 = 192 days. But we want the wall to be built in 16 days, 1/12 of 192 days. So we will need 12 times as many men as can do the job in 192 days, that is 12 times 1 man or **12 men**.

Symbolically, Let M represent men, D represent days, and W represent rods of wall. Then W/DM = constant. So W/DM = 50/(12)(10) = 80/M(16) = 5/12. So 5/M = 5/12 or the number of M is 12.

Section 7.3

1. Thirty squares should be shaded in the 10×10 grid.

3. 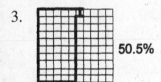 **50.5%**

5. 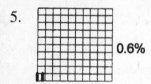 **0.6%**

7. **45%**   45/100 or 9/20   0.45

9. **175%**   7/4   **1.75**

11. 37.5%   37.5/100 or 3/8   0.375

13. 2.5%   2.5/100 or 1/40   **0.025**

15. The increase is 2000 with a base of 1000. So % increase is $\dfrac{2000}{1000} \times 100\% = 200\%$.

17. From 480 to 2400 is an increase of 1920 with a base of 480. So the percent increase is $\dfrac{1920}{480} \times 100 = 400\%$.

19. From 600 to 200 is a decrease of 400 with respect to a base of 600. So % decrease is $\dfrac{400}{600} \times 100\% = 67\%$.

21. 60% = 0.6 So 0.6(20,000) = 12,000 students are accepted. 40% = 0.4. Thus 0.4(12,000) = **4800 students enroll.** Note that this could have been done: let E represent students enrolled. Then $E = (.6)(.4)(20,000) = 4800$.

23. Let R represent the number of readers of the paper. Then we have $(4\%)R = 50$, $0.04\ R = 50$, $R = 1250$.

25. **Luis:** $49/59 = 0.83 = $**83%**. Felipe: $58/72 = 0.81 = 81\%$. Maria: $74/92 = 0.80 = 80\%$.

27. The percent increase was $[(20,000 - 8,000)/8,000]100\% =$**150%**.

29. The percent change in asking price was: $[(200,000 - 180,000)/200000]100\% = 10\%$ drop in price.

31. The increase is $1 with a base of $2.50. $\frac{1}{25}(100\%) = 40\%$ increase.

33. Answers will vary. 10% or 11% would be good estimates.

35. The use of the % key may vary among calculators. One possibility is to use the % key in place of the = key. The % key performs 2 functions. It first divides the product by 100 and then displays the product.
a. **19.2**   b. **161.28**   c. **0.625**   d. **577.5**

37. Estimates will vary. The tip is about 15% of $40 which is about 1/ 10 of 40 plus half that amount or $4 + 2 = \$6.00$. Essentially this is the reverse use of the distributive law: $15\%(40) = [10\% + 1/2(10\%)]\ 40 = (1/10)(40) + 1/2\ [(1/10)(40)]$.

39. Applying the same logic as in problem 33 we see that 1% of the number of bowlers is 1/100 of 725 or about 7 people. Another ½ percent is ½ of 7 or about 4 people. Thus we can expect about $7 + 4 = 11$ out of the 725 to bowl 200 or better.

41. Yes, a % decrease can be less than 1%. Consider a decrease from 200 to 199, a decrease of 1 with respect to 200. The percent decrease is $\frac{1}{200} \times 100\% = .5\%$.

43. The value of the stock will be **less** than the initial price. Arguments will vary. One possibility is: Let S represent the initial price of the stock. Then after the 15% drop the stock is worth $85\%\ S$. Now, after the 15% increase in price the value is $85\%S + (15\%)(85\%\ S) = 0.85\ S + 0.1275\ S = 0.9775\ S$ which is less than S.

45. Explanations will vary. Scenario 1: The net price of the TV is $85\%\ (600) = 0.85(600) = \$510$. The 6% tax is 6% of $510 or $0.06(\$510) = \$30.60$. So the total price, including tax, is $\$510 + \$30.60 = \$540.60$. In scenario 2 the total cost is: $(85\%)\ [\$600 + 6\%(\$600)] = 0.85\ [600 + 0.06(600)] = 0.85(636) = \$540.60$. **The costs are the same**. In general terms, letting C represent the original cost, D the percent discount, T the percent tax, and R the cost to Renaldo, we have in scenario 1, $R = [(1 - D)C](1 + T)$ and in scenario 2, $R = (1 + T)C(1 - D)$. The only difference is in the order of the multiplications. But multiplication is commutative and associative.

47. If O represents the original value, the value, V, after a percent, P, growth can be calculated as: $V = O + (P)(O) = O(1 + P)$. So, If P, Q, and R represent consecutive growths, the value after all three growths is: $V = \{[O(1 + P)](1 + Q)\}(1 + R)$. So, because multiplication is associative and commutative, scenarios a and b are equivalent. Now, for a and b $V = O(1.05)(1.10)(1.15) = 1.32825$ O whereas in scenario the value would be: $V = O(1.1)(1.1)(1.1) = 1.331$. So c **is the better investment.**

49. Suppose that 100 acres were lost to fire last year. Then 200 acres would be lost this year. The percent increase is $(200 - 100)/100 = 1$. So the increase is 100%, not 200%.

51. Suppose N represents the original number and P the percent decrease. The 5% increase gives a value of $1.05N$. Now we want to decrease this value by $P\%$ so that the new value $(1 - P/100)(1.05N) = N$. So we have $(1 - P/100)(1.05) = 1$; $1 - P/100 = 1/1.05$; $P = 100(1 - 1/1.05) = $ **4.76%.**

53. Increasing a number by 100% means doubling the number. So to return to the original value the new value must be halved. Taking **50%** of a number halves that number.

55. To arrive at a conjecture try some values for a and b. Because the order of multiplication does not affect the net raise we need not consider variations in order, first the smaller, then the larger and vice versa. It is good practice to consider three pairs for a and b: one in which the values are close, one in which they are equal, and one in which they are far apart. So first consider raises of 1% and 1 1/2%. The average is 1 1/4 %. For a $100 salary the separate raises would result in a salary of $100(1.01)(1.015) = \$102.515$. Applying the average twice we get $100(1.0125)(1.0125) = \$102.5156$. If the raises are equal, say 6%, the average is also 6% and the computations are the same. If the raises are quite different, say 1% and 20% we have an average of 10.5%. The separate raises give a salary of $100(1.01)(1.20) = \$121.2$ and the average, twice applied, gives: 122.10, a more substantial raise. So, **if the raises are not equal, the average twice applied gives a greater raise.**

57. Responses will vary among students.

59. Suppose C represents the number of couples. Then $0.52\,C - 0.39\,C = 100$. So $0.13\,C = 100$ and C is about **770 couples**.

SECTION 7.4

1. $(0.05)(\$2000) = \$100$

3. $(0.065)(\$10000)(2) = \$1300$

5. $(0.045)(\$8000)(1.5) = \$540.$

7. Let x represent the interest rate. The $(\$5000)(4)x = \$1400$; $x = 0.07$ or 7%

9. $a = \$4000 \left(1 + \dfrac{0.06}{365}\right)^{365(4)} = \$5084.90,$ so the interest is $1084.90

11. The 9%/18% interest rate plan is a better deal for the first 18 months. After that, the steady rate of 15% is a better deal.

Month	Charges	Interest Co. A	Total interest	Interest Co. B	Total interest
1	1000	90	90	150	150
2	1000	90	180	150	300
3	1000	90	270	150	450
4	1000	90	360	150	600
5	1000	90	450	150	750
6	1000	90	540	150	900
7	1000	180	720	150	1050
8	1000	180	900	150	1200
9	1000	180	1080	150	1350
10	1000	180	1260	150	1500
11	1000	180	1440	150	1650
12	1000	180	1620	150	1800
13	1000	180	1800	150	1950
14	1000	180	1980	150	2100
15	1000	180	2160	150	2250
16	1000	180	2340	150	2400
17	1000	180	2520	150	2550
18	1000	180	2700	150	2700
19	1000	180	2880	150	2850
20	1000	180	3060	150	3000

13. The display for $1000 invested at 6% compounded weekly for 3 years might be similar to:

End of year number	Value of investment
1	1061.80
2	1127.42
3	1196.09

If the years are in cells A2, A3, and A4 and the values are in cells B2, B3, and B4 then the formulas in the BX cells are 1000*(1+0.06/52)^(52*AX). Different spreadsheets have different methods of including this relative addressing into the formulas without direct entry from the keyboard.

15. The value, v of an investment, p, with annual interest r compounded n times yearly for t years is: $v = p\,(1+r/n)^{nt}$. So $v = 1000(1+0.08/365)^{(365)1} = \textbf{\$1083.28.}$ Since $1000(1+0.08328) = 1083.28$ the simple interest would be **8.328%**.

17. The value, v, of an investment, p, with annual interest r compounded n times yearly for t years is: $v = p(1 + r + n)^{nt}$. Since compounding is yearly, $n = 1$, $p = 1000$, and $v = 1000(1 + r)^t$

Yearly interest rates		6%	8%	12%
	Year			
a.	6	1419	1587	1974
b.	7	1504	1714	2211
c.	8	1594	1851	2476
d.	9	1689	1999	2773
e.	10	1791	2159	3106
f.	11	1898	2332	3479
g.	12	2012	2518	3896
h.	18	2854	3996	7690

19. If the depreciation is about 10% per year then the value at the end of a year is about 90% of the value at the beginning of a year. So after 5 years of depreciation at 10% per year the value is $(.9)(.9)(.9)(.9)(.9)(\$20{,}000) = \mathbf{\$11{,}810.}$

21. At the end of 10 years Rudy had (10 year)(12) month/year)($100/month) + $1000 = $13,000. This money is now invested for 30 years at 10%, compounded yearly. So the total amount at the end of the 30 year investment period is $13000(1.1)^{30} = \mathbf{\$226{,}842.23}$.

23. Suppose that Mark has 100 dollars to invest at 5% *simple* interest. Then he makes $5 per year. If the rate doubles to 10%, he makes $10 per year, double what he made at the lower rate. Now, Suppose Monica invests 100 at 5% *compounded* quarterly. Then in the first year she makes $100(1 + .05/4)^4 = \$105.09$. If the rates double to 10% she makes $100(1 + 0.010/4)^4 = 110.38$ which is more than double her earnings at the lower interest rate. Mark is correct for simple interest, Monica for compound interest.

25. If the number of denarii doubles every 5 years and there are 20 sets of 5 years in 100 years, there will be $2^{20} - 1 = 1{,}048{,}575$ denarii after 100 years. You can check this using a spreadsheet or a calculator. To determine the rate of compound interest, solve

$$1{,}048{,}575 = 1\left(1 + \frac{i}{365}\right)^{365(100)} ; 1{,}048{,}575 = \left(1 + \frac{i}{365}\right)^{36500} ; 1.00379879 = 1 + \frac{i}{365} ; i = 13.9\%.$$

CHAPTER 7 REVIEW EXERCISES

1. Responses will vary. Possibilities include:
    a. A whole to part ratio is the ratio between all the students in the sixth grade class, S, and those students in the class who have a parent who speaks French at least occasionally, F. $S{:}F = 25{:}8$.
    b. A part to part ratio is the ratio between F and those students in the class with parents who speak no French, N. This ratio is $F{:}N = 8{:}17$.
    c. A part to whole ratio is the ratio between N and S. $N{:}S = 17{:}25$.

3. Converting to unit prices we have: a. $1.65 per 10 oz = 16.5 cents/oz; b. $1.50 per 9 oz =16.7 cents/oz; c. $.96 per 6 oz = 16 cents/oz. So the order, best price to worst, is **c, a, b.**

5.

p	2	3	4	5	6
q	1.5	2.25	3	3.75	4.5

7. a. **True.** This is one of the definitions of proportional variation.
    b. **False.** Suppose $k = 4$. Then if $s = 2$ $r = 6$ and if $s = 4$, $r = 8$. 2/6 is not equal to 4/8. So s and r are not proportional.
    c. **False.** Let $r = 8$ and $s = 4$. Then $r/s = 8/4 = 2/1$. And $(r + 2)/(s + 2) = 10/6 = 5/3$. Since 2/1 is not equal to 5/3, this counter example shows that, in general r/s is not equal to $(r + 2)/(s + 2)$.
    d. **True.** This is the fundamental law of fractions and the numerators and denominators of equivalent representations of the same numbers are proportional.

9. a. $(x - 2)/12 = 5/6$; $(x - 2)/12 = 10/12$; so $(x - 2) = 10$ and $x = 12$.
    b. $.4/5 = x/7.5$; $5x = (.4)(7.5) = 3$; $x = \mathbf{0.6}$.
    c. $15/4 = 45/x$; $15x = 4(45)$; $x = \mathbf{12}$.
    d $4/x = 21/56$; $21x = 4(56)$ $x = \mathbf{10\ 2/3}$.

11. Each ball costs $3. Three balls would cost $9.

13. | Percent | Decimal | Fraction |
      |---------|---------|----------|
      | **1/2 %** | 0.005 | 1/200 |
      | 350% | **3.50** | 350/100 or 7/2 |
      | 4.5% | 0.045 | **9/200** |

15. Let S represent seatbelt users. Then $0.87(1700) = S = 1479$.

17. The percent decrease is: $(29,500 - 22,000)/29,500 = 25.4\%$.

19. Interest = P · R · T. So interest = ($2200)(.055)(2) = $242.

21. The methods of estimation and the estimates will vary. One possibility is: The room cost for 2 nights is $170. Together the taxes are 18%. Overestimate the taxes as 20% and underestimate the room cost as $150 dollars and the tax is estimated as $30. Add this to the actual room cost and the total comes to about $200.

23. Of the 1000 people contacted 500 were men and 500 women. Half the men, 250, said they would use the product. One fifth of the women, 100, said they would use the product. Thus 150 more men than women said they would use the product.

25. Of the initial 600 people 3 of 5 or 60% will respond. So 360 respond, 240 do not. Of the 240 one of three respond on the second mailing. Thus an additional 80 people respond leaving 160 who did not respond on either the first or second mailing. Of these, one in 8, or 20, can be expected to respond on a third mailing. So of the 600, 460 can be expected to respond over three mailings.

27. 1% of a = (l/1000) a = a/1000. 200.5% of b = (200.5/100)b = 2005b/1000 = 401b/200.

29. Since the value, v, of an investment, p, with annual interest r compounded n times yearly for t years is: v = p(1 + r/n)nt, with a rate of 5% compounded daily and for a time of 1 year, some principle p would have a value of v = p(1 + 0.05/365)1(365) = p(1.0513). Since for simple interest we have v = pit and the time is 1 year, the simple interest equivalent is 5.13%.

31. Discussions will vary. A 1 by 10 grid would serve as a model rather than our 10 by 10 grid. 0.5% would be 0.5/10 and 5% would be 5/10. A five percent increase would be half again as much. A 50% increase would mean five times as much as the original.

SECTION 8.1

1.   a.-b.

Word Length	Frequency	Relative Frequency
2	6	12
3	10	20
4	5	10
5	12	24
6	8	16
7	6	12
8	0	0
9	1	2
10	2	4

c. Dot Plot of Word Lengths

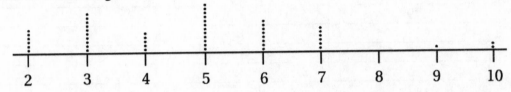

d.   Yes, the word lengths of 9 and 10 letters. These were longer and separated from other values by a gap of length 1.

3. a.-b.

Death Age	Frequency	Relative Frequency
98	2	4
93	1	2
90	2	4
89	3	6
88	1	2
87	1	2
86	3	6
85	1	2
83	1	2
82	1	2
81	1	2
80	1	2
79	3	6
78	1	2
76	3	6
75	3	6
74	2	4
73	1	2
72	1	2
70	3	6
69	1	2
68	1	2
66	2	4
64	1	2
59	1	2
58	1	2
57	1	4
56	1	2
55	1	2
53	1	2
51	1	2
48	1	2
46	1	2
41	1	2

c.
```
4 | 168
5 | 1356789
6 | 46689
7 | 00023445556668999
8 | 0123566678999
9 | 00388
```
5|1 **means** 51 **years of age**

d.

Age at Death for Rural Sample

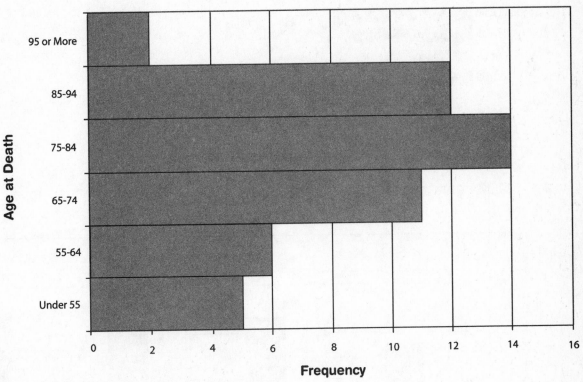

e.   The 41-46 ages and the two 98 year-olds.

5.
```
A| ••••••••••
B| •••••••••
C| •••••••
D| •••
F| •
```

7.   **Distribution of Grades in Ike's Class**

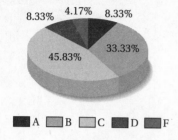

9.   **Most Popular Girls High School Sports**

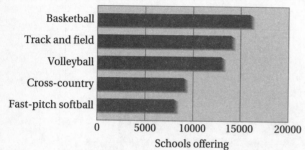

11. a.

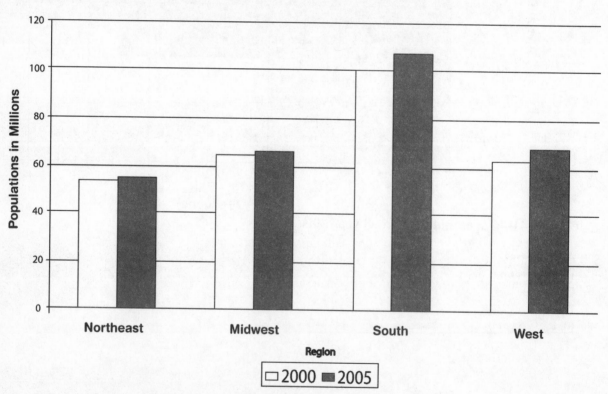

b.   the west grew most rapidly in the period by 7.6%. Its growth was followed closely by the south with a 6.9% growth.

13. a.   199 | 1788899
         200 | 0112334444556   199|8 **means** 1998

   b.   15 years
   c.   2004

15. The students in the class have scores from 60 to 100. The scores are spread out with 3 getting 100s, and several students from 99 to 91. Then there are students every 2 to 3 points on the scale until a cluster at 72 to 74.

17. The number of hours Americans work weekly varies considerably from 1 hour to those that work over 49 hours per week. The greatest percentage, 41.5%, work the traditional 40-hour work week. Nearly 31% of the working population works less hours than 40, while 27.6 work more. Beyond those working 40 hours a week, the largest segment of the working population is the 17.7% working more than 49 hours per week.

19. This circular graph depicts the partitioning of the population of office workers into whether or not it is easier to balance responsibilities today than five years ago. The results indicate that nearly 3/5 of the respondents feel that it is not easier to balance the home-work responsibilities today than it was five years ago. On the other hand, about 1/3 indicated that it was easier. Another 11 percent were not sure or felt that were no differences. Overall, the graph depicts a population of office workers feeling that they are being pulled in different directions harder today than five years ago.

21. Responses will vary. Students may select a bar graph because the categories are non-numeric. A possible graph is:

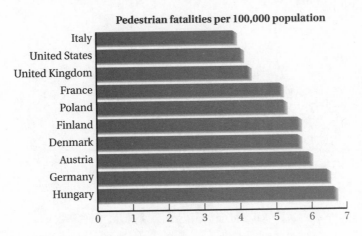

Pedestrian fatalities per 100,000 population

23. Since the data represent the total numbers of students served by federal programs, one way of representing it is to show the proportions each type of program accounts for within the disabled pool of students. The graphic would look somewhat like that shown below. The number of categories is large and may make interpretation difficult. Another way would be to use a bat graph.

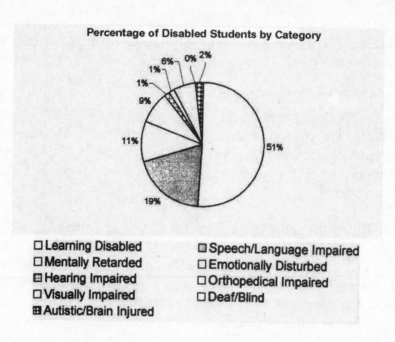

25. Responses will vary. Because of the number of categories, a bar graph is appropriate. A representative graph is:

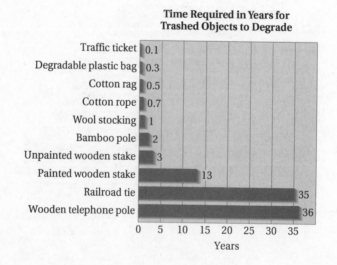

27. a. The group with the highest AIDS deaths in 2004 was the 40-49 age population.
    b. Rate of exposure in group before risk of AIDS was known and time to move from HIV positive to AIDS.
    c. The rate increases up to and including the 40-49 age group then decreases.
    d. The 40-49 age group was the first with large exposure rate and education has helped decrease probability of protraction since that point. Risk was less prior to this group.

29. **Dollars in Circulation (in trillions)**

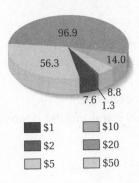

96.9
56.3
14.0
8.8
7.6
1.3

■ $1   ☐ $10
■ $2   ▨ $20
☐ $5   ☐ $50

31. (a) **Number of Suicide Deaths in the United States, 1989**

Females	Age	Males

| Under 15 |
| 15–24 |
| 25–34 |
| 35–44 |
| 45–54 |
| 55–64 |
| 65–74 |
| 75–84 |
| 85 & Older |

4000 3000 2000 1000   1000 2000 3000 4000 5000 6000

(b) For each age, more men than women commit suicide. Both peak in the 25–34 age range.

(c) As before, more men than women at each age, but the ratio of men to women is fairly constant.

33. (a) **Comparision of Test Scores for Class A and Class B**

Class A		Class B
100	14	01112
552100	13	00122578
987650	12	1678
985	11	8

7|14 means 147   14|7 means 147

(b) Class B had slightly higher performance on the test.

(c) In class A, outliers appeared at 115 and 140–141, with clusters at 115–120, 125–132, and 140–141; in class B, outliers appeared at 118–121, with clusters at 126–132 and 137–142.

35. Responses will vary. Some responses might be similar to: A dot plot is generally used when specific values are repeated. A frequency plot is used when the values are placed in categories. The dot plot is used for, generally, fewer values than the frequency table. A dot plot might be used to display the scores on a quiz of, perhaps, 10 points for a class of 15 to 30 students. A frequency table might be used to display the scores of a lecture class, maybe 200 students, on an exam of 200 points.

37. The graph and information indicates that 7 out of 10 adults send cards, but that the vast majority of these, 78 percent, send fewer than 50 cards. Only fiver percent send more than 100 cards. There is a marked drop in card sending at the 50 card per year level from the previous levels.

SECTION 8.2

1. Responses may vary. A possibility: Except for a 1% decrease in 1994, projected costs show increases for each year from 1993 to 1998. Following a large increase in 1993 and the '94 decrease, the increases become larger.

3.

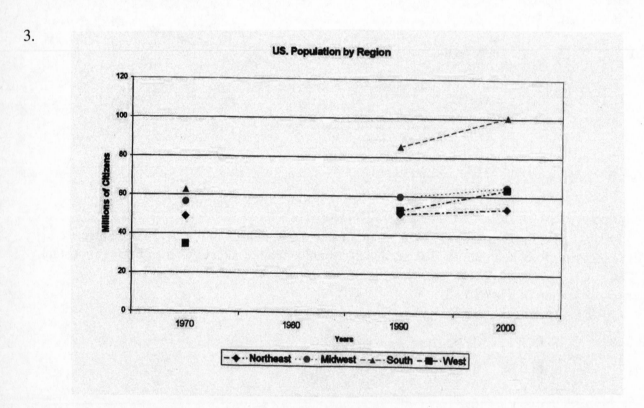

5. Between 1973 and 2005 students' scores have increased approximately two-grade levels in mathematics. These increases occurred in two major periods, one grade-placement occurred between 1986 and 1990 and the second between 2000 and 2005..

7. (a) shows no correlation between the variables. (b) shows a relatively strong positive correlation and (c) shows a negative correlation. (d) a relatively weak positive correlation.

9. Responses will vary. (b), with the long vertical axis, emphasizes small changes in the vertical variable.

11. No, the trend line would not be appropriated to give an expected class average for an ACT of 34. The graph does not continue linearly because there are maxima to both ACT and class average. The extension of the trend line gives a class average greater than 100 for an ACT score of 34.

13. a. There is a positive correlation between the variables. In general, an increase in the value of one variable is accompanied by an increase in the value of the other variable.
    b. By reducing the separation of the points on the vertical axis the correlation could be made to appear stronger. This could be done by increasing the maximum value of the window in the vertical direction from 60 to, perhaps, 100.

15. Responses will vary. One approach is: Because the emphasis is on the growth of population, not the actual population figures, the data will be best represented with a bar graph.

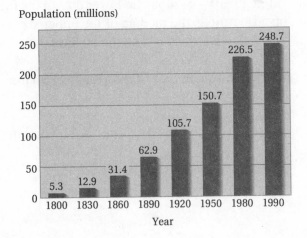

17. Responses will vary with the research of the students.

SECTION 8.3

1–5 Before performing computations, order those data sets that are not initially ordered. The sum of each set is in parentheses after the set. The middle value, or the middle pair, of each data set is underlined.

   0 0 1 2 2 4 5 10 (24)   6 7 8 9 10 12 15 15 20 28 (130)   72 80 80 82 88 90 96 (588)
   61 68 68 73 85 91 93 (539)   68 74 80 82 83 85 86 88 91 93 (830)   –15 –2 –1 0 0 7 7 8 14 (18)

   To determine the mean divide the sum of the data by the number of values. The median is either the middle value of the mean of the middle pair, the mode is the most frequently occurring value, and the midrange is the mean of the highest and lowest values.

	mean	median	mode	midrange
1.	**3**	**2**	**0, 2**	**5**
2.	**13**	**11**	**15**	**17**
3.	**84**	**82**	**80**	**84**
4.	**77**	**73**	**68**	**77**
5.	**83**	**84**	**None**	**80.5**
6.	**2**	**0**	**0, 7**	**–.5**

7. The mean salary is [49(25,000) + 125,000]/50 = **27,000**. Both median and mode are **25,000**. If the largest salary is dropped from the data then the mean is [49(25,000)]/49 = **25,000**. The median and mode are unchanged.

9. The ordered data set is: 5 6 7 7 8 8 8 9 9 9 10 10. The **range is** (10 – 5) = **5**. The lower quartile is the median of the lower six scores, (7 + 7)/2 = 7 and the upper quartile is the median of the upper six scores, (9 + 9)/2 = 9. So the **interquartile range** is (9 – 7) = **2**. The mean is 8 so the **variance** is $[(5 – 8)^2 + (6 – 8)^2 + (7 – 8)^2 + (7 – 8)^2 + (8 – 8)^2 + (8 – 8)^2 + (8 – 8)^2 + (9 – 8)^2 + (9 – 8)^2 + (9 – 8)^2 + (10 – 8)^2 + (10 – 8)^2]/12 = $**2.166**. The standard deviation is the square root of the variance. So the **standard deviation is 1.472**.

11. The **range** is (28 – 6) = **22** and the **interquartile range** is (15 – 8) = 7. The **variance is 41.8** and the **standard deviation is 6.47**.

13.
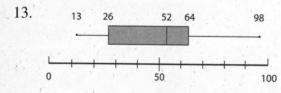

15.
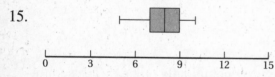

17.
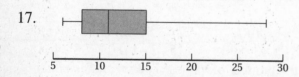

19. Responses will vary. They may include statements similar to: Network *A* has both the highest and lowest rated shows. Nearly half of the shows on *A* have ratings equal to or better than the top quarter of the shows on *B*. The median rating of shows on *A* almost matches the rating of the third quartile of the shows on *B* and the third quartile of the ratings of the shows on *A* is almost the same as the highest rated show on *B*.

21. Since the mean is the sum of the data divided by the number of data, the sum of the data is the product of the mean and the number of data. Thus the sum of the 4 scores is 4(72) = 288.

23. The sum of the 25 scores is 25(80) = 2000. The sum of all 27 scores is 2000 + 30 + 35 = 2065. Thus the mean of the 27 scores is 2065/27 or, as a decimal, 76.481.

25. a.

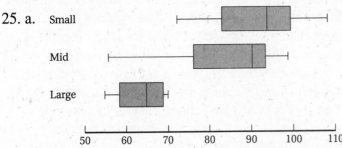

   b. Responses will vary. They may include statements similar to: Large cars have, by far, the least injury potential. All of the large cars have a lower injury potential than the safest quarter of the midsized and small cars. And all of the midsized cars are potentially safer than all but the safest 25% of the small cars.

27. Average speed, *s*, is equal to the total distance, *d*, divided by the total time, *t*: $s = d/t$. Since Steve drove at 30 mi/hr for 2 miles it took him 2 mi/30 mi/hr = 1/15 hr to drive the first 2 miles. In order to have an average speed of 60 mi/hr for the entire 4 miles trip he must make the entire trip in 4 mi/60 mi/hr = 1/15 hr. But he has already taken 1/15 hr for the first 2 miles. So no matter how fast he drives on the second 2 miles, he cannot attain an average speed of 60 mi/hr.

29. It is possible for the mean of a set of data to fall outside the box-and-whisker box. Consider the data set:

   L *Q*1   *M*   *Q*3   *H*
        2.5   7.5
   1 2 3 4 5 6 7 8 *X*

   We can assign any value greater than 8 to *X* and not affect *Q*1 and *Q*3 which determine the size and location of the box. If we want the mean outside of the box then we want the mean to be greater than 7.5. Thus we want the total of the data to exceed 9(7.5) = 67.5. The data total, without *X*, is 36. So for any value of *X* greater than 31.5 the mean will exceed 7.5.

31.

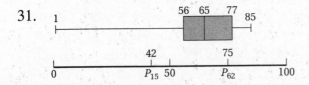

The 15th percentile score is 42 because 15% of the scores fall below that score. The 62nd percentile score is 75 because 62% of the scores fall below that one. The values for other measures are $Q_1 = 56$, median = 65, and $Q_2 = 77$.

33. **NO**, the standard deviation is not always less than the variance. Suppose that the variance is a number between 0 and 1. Then the square root of the variance, the standard deviation is greater than the variance. For example, suppose the variance were 0.81. Then the standard deviation would be 0.9.

35. Suppose that the original data are a, b, c, and d. Then the mean is $(a + b + c + d)/4$. Now, if 10 points are added to each of the scores, then the mean becomes $(a + 10 + b + 10 + c + 10 + d + 10)/4$ $= (a + b + c + d)/4 + 10$. So the mean has increased by 10 points. Because the mode is one (or more) of the data points, adding 10 points to each value would increase the mode by 10 points. If the median is one of the data (there are an odd number of points) then the median would be increased by 10 points because all data are increased by 10 points. If there are an even number of data, then the median is a mean of 2 points and, as in the case of the overall mean, would increase by 10 points. The standard deviation is calculated from the differences between each value and the mean. Since each of these is increased by 10 points the differences are unaffected and the standard deviation is not affected by the addition of the 10 points to each score.

37. The original data set, O, {0, 12, 13, 15, 17, 17, 18, 20} becomes $N = \{5, 12, 13, 15, 17, 17, 18, 25\}$. Because the mean is the sum of the data values divided by the number of values and since the sum has increased but the number of values has not, the mean will increase by 10/number of values. The median will not be affected because its determination does not involve H and L. Since neither H nor L is the mode, the mode will be unaffected. The midrange will increase by 5. The median of both sets of data is 16. The mode of both sets is 17. The midrange of O is $(0 + 20)/2 = 10$, the midrange of N is $(5 + 25)/2 = 15$. The mean of O is 14, of N, 15.25.

39. The mean, median, and midrange need not be members of the data set that they describe. The are calculated from members of the data set. The median is always a member of the set for an odd number of data. The mode is always a member of the data set because it is that member of the data most frequently occurring.

41. Responses will vary depending upon the definition of 'middle' of a data set. Some students may chose the median and argue that the middle is more related to the number of data than to the values of the data. Others may use the 'balance point' idea and chose the mean as the best representative of middle.

43. Responses will vary depending upon the research of the students.

45. Responses will vary among students. They may discuss that variance and the related standard deviation have most meaning for normally distributed data. The standard deviation is related to the probabilities of outcomes in some defined range of the possible outcomes. These relations are of interest to those who participate in games of chance. Initially they may help one determine if a game is 'worth' playing.

47.

X2 Mean	Y2 Mean	X2 Standard deviation	Y2 Standard Deviation
9.00	7.50	3.32	2.03

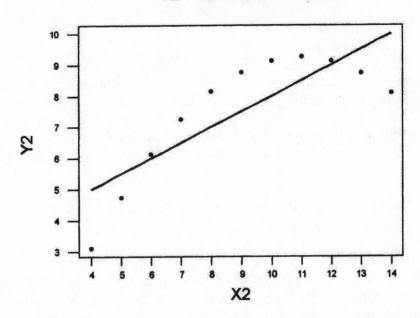

49.

X4 Mean	Y4 Mean	X4 Standard deviation	Y4 Standard Deviation
9.00	7.50	3.32	2.03

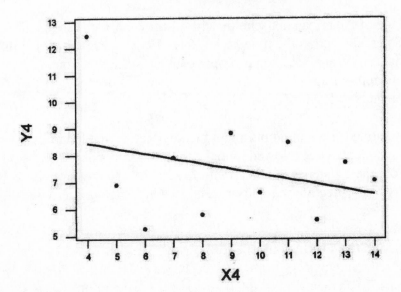

50. Note that the means and standard deviations are the same for the X, or independent variables, and Y, or dependent variables. However, the scatterplots and linear trend lines are quite different. In fact, some data sets would be better fit by curved trend lines.

SECTION 8.4

1–5.   Responses will vary. Some considerations are:

1. The statement does not describe either the characteristics of the sample of persons responding or the criteria upon which they were to make judgements of the greatest American jazz musician.

3. The sample of persons surveyed may very well not be representative of the snacking American population. Further, they may have been responding to a perceived question: "What is your favorite snack at a ball game?".

5. There are many levels of professional baseball, class D through the majors. The statements does not define the population of player to which the conclusion is to be applied.

7. Responses will vary. Some considerations are: All the barrels have the same diameter (variations are related to perspective, not to actual size). Thus the volume differences would be first power functions of the heights. But the graphic shows the increase in the cost per barrel-the volume of oil associated with each price is the same. If the symbols in the ad were to be retained, the barrels should be the same size and the price tags should be sized in area in proportion to the price of one barrel of oil.

9. Responses will vary. Considerations are: The stem and leaf presentation supports those promoting Ruth as the greatest home run hitter. Three quarters of Ruth's seasons were better than all but 3 of Maris' seasons.

11. a. The graph shows that the percentage of American households with insufficient food remained fairly constant, from 10.1 to 11.8 percent, across the time period from 1998 to 2003. It was at its highest in 1998, then dropped to its lowest and has been creeping back up since that time.
    b. The graphic display chosen is quite appropriate.
    c. Answers will vary—perhaps a line graph with dotted trend lines.

13. a. The graph shows the results of an online survey of adults about the key factors that led to their purchase of some technology. Respondents were allowed to make multiple responses. The results show that major factors ranged from the Existence of a Warranty for 50% of the buyers to Ease of use for 61 percent of those surveyed. Three of the top four responses indicated concern over use and start-up issues.
    b. Graph format is appropriate.
    c. One might also have used a pictograph to show the information.

15. The total minority enrollment has steadily increased over the last 20 years both in absolute numbers, 948,00 to over 2 million, and as a percent of total college enrollment, 10% to almost 16% minority enrollment as percent of total enrollment

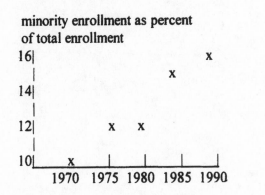

minority enrollment as percent of total enrollment

```
16|                        x
  |                   x
14|
  |
12|         x    x
  |
10|    x    |    |    |    |
      1970 1975 1980 1985 1990
```

Minority enrollment (Millions)

```
2.1|                              x
2.0|
1.9|
1.8|
1.7|                        x
1.6|
1.5|                   x
1.4|
1.3|              x
1.2|
1.1|
1.0|
0.9|    x    |    |    |    |
      1970 1975 1980 1985 1990
```

17. Responses will vary. Some arguments may hold that there is a point of diminishing returns, possibly negative returns, related to study time. Some students may not be able to match the criteria for high grades no matter how long they study. There is a correlation, perhaps even a causation, between study time and grades but it is not the same for all students.

19. Responses will vary. One approach is: As you talk to the personnel of Department A: "Last year your department proved so valuable to the company that we have decided to increase your funding by a full 5% so that you can not only maintain the level of your accomplishments but will also have the funds to go even further. We are only going to increase the funding for Department B in proportion to the overall growth of the company. But you, Department A, will receive a full 5% increase."

CHAPTER 8 REVIEW EXERCISES

1. a.   Mean = 5.448; median = 5.46; modes = 5.29 and 5.34; range = 0.97.

    b.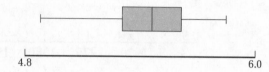

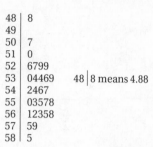

```
48 | 8
49 |
50 | 7
51 | 0
52 | 6799
53 | 04469      48│8 means 4.88
54 | 2467
55 | 03578
56 | 12358
57 | 59
58 | 5
```

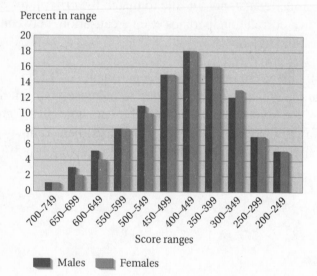

3. a.

b.

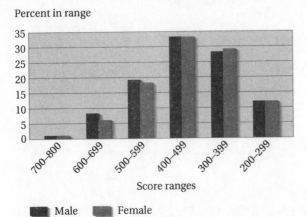

Male and Female Performance by Range

Male   Female

This graph shows less difference in the performance by gender per score range.

5. a. The data set for 'Sightings Confirmed' is {2, 2, 3, 3, 4, 5, 5, 8, 9, 10, 12, 13, 13, 14, 16} and for 'Reported Sightings' is {2, 3, 3, 3, 4, 5, 6, 9, 10, 11, 13, 15, 15, 16, 19}. Because there are 15 data values in each set the median values are the 8th counting from lowest: **8 Confirmed Sightings and 9 Reported Sightings**.

   b. The graph suggests a **positive correlation** between reported and confirmed sightings because as one increases the other increases.

   c. A trend line would not be appropriate because the number of data is close to the tornado population for any May. Extending the data to larger numbers of tornados is misleading.

7. The conclusion on the city's "best" restaurant is not valid. The criteria for best is not defined. The crowd at a ball game may not be a sample representative of the population as a whole and the first 50 may be snack bar aficionados who planned to eat at the snack bar because they consider that to be fine dinning.

9.

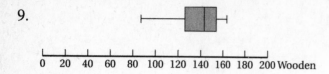

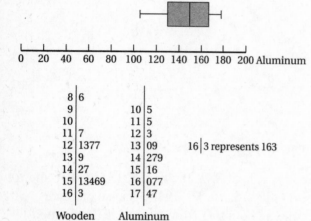

```
 8 6
 9          10 5
10          11 5
11 7        12 3
12 1377     13 09        16│3 represents 163
13 9        14 279
14 27       15 16
15 13469    16 077
16 3        17 47
```

    Wooden   Aluminum

Arguments will vary. They may include the concepts that the aluminum bat provided more consistency as well as greater distance hitting.

11.   **Reading Scores**

```
5│1
4│0012456
3│135589        3│1 represents a score of 31
2│689
1│445
```

Number of scores

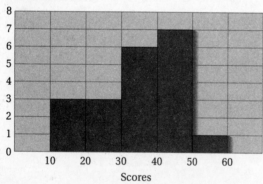

Scores

Responses will vary. The histogram conveys the primary information, distribution of, and general values of, the reading scores in a more emphatic manner.

13. Select a random sample of individuals, balanced for right- and left-handers and by gender; use a set of hand-exercise grips equipped to measure the foot-pounds of force exerted by each hand; graph and chart the results individual by individual; analyze the results.

15. Responses will vary but should include consideration of sample size, sample selection, survey design, and data analysis and display.

17. The data on patient temperatures is perhaps best displayed by a stem-and-leaf plot or, given the large number of temperatures, a box-and-whisker plot. Doing the former, we have the stem-and-leaf plot shown:

```
 96 | 7
 97 | 348899
 98 | 5556666666667789
 99 | 0379
100 | 3
101 | 7
102 | 1447
103 | 2668
104 | 18
105 | 2          105|2 means a temperature of 105.2
```

These data show that the set of data is bunched about normal temperature and a degree or two to each side of this value—98.6. Given that there are more outlying temperatures above this cluster, it is probably better to use the median temperature in a written report. An examination shows a mean of 99.97, while the median is 98.7. Here one can see the effect the higher outlying temperatures have on the reporting of a central value.

19. Calculating the percent of the total Jaytown students attending each of the three high schools, we get 36 percent at Jim High, 24.6666 percent at Jean High, and 39.3333 percent at Jan High. Taking these representatives percentages of the 150 students to be sampled, we get the sample sizes of 54 for Jim High, 37 for Jean High, and 59 for Jan High. Within each of the high schools this same process could be repeated to partition the school's sample size among the classes to get a truly representative sample. The, the number of students determined should be randomly drawn by lots from the respective student groups they would represent.

SECTION 9.1

1. Since every element of the set $\{m, a, t, h\}$ is also an element of the set $\{m, a, t, h, e, i, c, s\}$, it is <u>certain</u> that if an element is chosen from the set $\{m, a, t, h\}$ then that element is also chosen from the set $\{m, a, t, h, e, i, c, s\}$. Thus the probability is **1**.

3–5.   The sample space of equally likely outcomes, S, for a fair die is $\{1, 2, 3, 4, 5, 6\}$.

3. For event $E = \{5, 6\}$, $P(E) = 2/6 = \mathbf{1/3}$.

5. For event $E = \{1, 2, 3, 4, 5\}$, $P(E) = \mathbf{5/6}$.

7. The primes in the sample space are 2, 3, 5. So the probability of rolling a prime is **3/6 or 1/2**.

9–11.   The sample space, S, contains 52 equally likely outcomes.

9. Considering the aces as face cards, E is a subset of S with 16 equally likely outcomes.
    So $P(E) = 16/52 = \mathbf{4/13}$.

11. Using the Addition Property of Probability (for non-mutually exclusive events) we have:
    A = non-face card, B = a seven; $P(A) = 36/52$, $P(B) = 4/52$, $P(A \text{ and } B) = 4/52$ and so $P(A \text{ or } B)$ is
    $36/52 + 4/52 - 4/52 = 36/52 = \mathbf{9/13}$. Note that when B is a proper subset of A, the probability of
    $(A \text{ or } B)$ is just the probability of A.

13. There are 4 sixes in the 52 card sample space. So the probability is **4/52 or 1/13**.

15. The multiples of four in the sample space are 4 and eight. There are eight of these so the probability
    is 8/52 or **2/13**.

17. $\dfrac{4}{36}$, the successes are 2, 3, 5, and 7.

19. $\dfrac{0}{36}$, there is no slip that contains both a digit and a letter on it.

21. There are 26 equally likely outcomes in the sample space. Of these, we will consider the vowels to
    be A, E, I, O, and U. Thus there are 21 consonants. Fifteen of the letters are seen to be formed from
    straight-line segments only and 7 have enclosed regions. So:
    a. P(consonant) = **21/26**.
    b. P(formed by straight-line segments only) = **15/26**.
    c. P(letter has enclosed region) = **7/26**.

23. Since A and B are complements, by definition of complement, $P(A) + P(B) = 1$. So $x + 0.7 = 1$
    and $x = \mathbf{0.3}$.

25. $\frac{3}{6}$, as the probability would be $1 - \text{P}(\text{odd}) = 1 - \frac{3}{6}$

27. $\frac{1}{6}$, as the only success is 2.

29. Since there are three outcomes in the sample space, $P(A) + P(B) + P(C) = 1$ because one of the events A, B, or C must happen. So since $P(A) = P(B)$ and $P(C) = 2P(A)$ we have $P(A) + P(A) + 2P(A) = 4P(A) = 1$ and **$P(A) = 1/4$**. So **$P(B) = 1/4$ and $P(C) = 1/2$**.

31. Using the data from the table
    a. $P(\text{prior}) = $ **46/80**.
    b. $P(\text{priors} > 2) = $ **6/80**.
    c. $P(\text{no prior}) = 34/80$, $P(\text{prior}) = 46/80$. So the probability of getting a student with a prior in-school suspension is **greater** than getting a student with no prior in-school suspensions.

33. Because there are 52 ways to draw the first card and, for each of these there are 51 ways to draw the second card, there are (52)(51), 2652, ways of drawing two cards. Similarly, there are 26 ways to first draw a red card and, for each of these, there are 25 ways to draw a second red. So there are (26)(25), 650, ways to draw two red cards. Thus the probability of drawing two red cards is 650/2652, **0.245**.

35. $P(A \text{ or } B) = P(A) + P(B) - P(A \text{ and } B)$. So $P(A \text{ and } B) = P(A) + P(B) - P(A \text{ or } B) = 1/2 + 1/6 - 1/2$ = **1/6**.

37. a. Suppose $P(A \text{ and } B \text{ and } C) = 0$. Then there are no elements common to all three of the event subsets of the sample space S. But there may be events common to some pair of these event subsets. Thus the elements of A, B, and C may not exhaust S. Consider the example $S = \{1, 2, 3, 4, 5, 6, 7, 8, 9, 10, 11, 12, 13\}$, $A = \{1, 2, 3, 4\}$, $B = \{3, 4, 5, 6, 7, 8, 9\}$, $C = \{10, 11\}$. $P(A) = 4/13$, $P(B) = 7/13$, $P(C) = 2/13$, $P(A \text{ and } B \text{ and } C) = 0$. So, we can say that $P(A \text{ or } B \text{ or } C) \leq 13/13$.
    b. But if the combination condition is changed to $P(A \text{ and } B) = P(A \text{ and } C) = P(B \text{ and } C) = 0$, then A, B, and C exhaust S. So we can say $P(A \text{ or } B \text{ or } C) = 13/13$.

39. The uniform sample space, S, for the 12-sided die could be $\{1, 1, 2, 2, 3, 3, 4, 4, 5, 5, 6, 6\}$ producing $P(1) = 2/12$ or $1/6$. $P(2)$, $P(3)$, $P(4)$, $P(5)$, and $P(6)$ are also $1/6$, the same as on a common 6-sided die.

41. $P(H \text{ Leclerc}) = 2048/4040 = $ **0.5069**. $P(H \text{ Pearson}) = 12012/24000 = $ **0.5005**. Theoretical probability estimates improve with larger sample sizes, so it appears that a better estimate would be obtained by a combination of the data. $P(H \text{ combined data}) = 14060/28040 = $ **0.5014**. There is an apparent anomaly here. If we believe that the 'real' probability of heads is 1/2, then adding Pearson on top of Leclerc seems to help while adding the Leclerc data on top of the Pearson data appears to hurt. Because Leclerc and Pearson were conducting different experiments, different coins, the data should not be combined. Leclerc's experiments estimate the theoretical probability associated with his coin, Pearson's experiments with *his* coin.

SECTION 9.2

1. Any one of the three letters may be placed in the first position. But for each of these only 2 letters may be placed in the second position. For each of the six of the placements of the first two letters only 1 letter may be placed in the third position. Thus there are $(3)(2)(1) =$ **6** different code symbols. These are: SPY, SYP, PSY, PYS, YSP, YPS.

3. There are 3 possible choices for the first character in the code, 2 possible choices for the second character in the code, and 10 possible choices for the third character in the code. Thus, there are $(3)(2)(10) = 60$ possible codes that can be made.

5. Any of the four cars may be parked in the first space. Any of the remaining three cars may be in the second space. Either of the remaining two cars may be in the third space and the last car must occupy the last space. Thus there are $(4)(3)(2)(1) =$ **24** different arrangements of the four cars.

7. For each of the six tops there are four bottoms. And for each of these 24 top/bottom combinations the child can select from five pairs of shoes. So there are $(6)(4)(5) =$ **120** different outfits.

9. We know that the first card drawn was a king and that this card is not to be replaced. So the deck now contains 51 cards, 3 of which are kings. The probability of drawing one of these kings is **3/51**. Because we know that the first card drawn was a king, the probability of drawing 2 kings is just the probability of getting a king on the second draw.

11. After the first two draws there are 50 cards remaining and, since no spades have yet been drawn, there are 13 remaining spades. Thus the probability is **13/50**.

13. The sample space for selecting a number from the set of natural numbers 1 through 9 is {1, 2, 3, 4, 5, 6, 7, 8, 9}. But if we know that the number selected is prime it is certain that the number selected is in the set {2, 3, 5, 7}. Of the 4 equally likely outcomes in this restricted sample space, 3 are odd. So the probability of selecting an odd number, knowing that it is prime, is **3/4, or 0.75**.

15. a. The probability of getting a blue ball on the first draw is 3/7 since 3 of the 7 balls are blue. Because the first ball was not replaced there are six balls remaining, two of which are blue. Thus the probability of getting blue on the second draw is 2/6. So the probability of drawing 2 blue balls is $(3/7)(2/6) =$ **6/42, or 1/7, approximately 0.14**.
    b. The probability that both are blue given that the first is blue is just the probability of drawing a blue on the second draw, **2/6 or 1/3**.
    c. The probability of white on draw 2, given blue on draw 1, is **4/6 or 2/3**.

17. The information available is displayed as: Now, since we know that the home has a detached garage, our restricted sample space is the shaded area. If we were dealing with 100 homes total, then the restriction of a detached garage limits the sample space to 67 homes of which 13 have patios. So the probability of patio given detached garage is **13/67**.

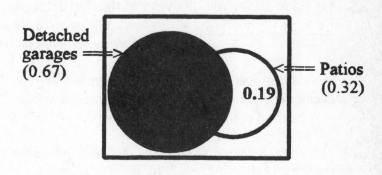

19. a.   {(B, B), (B, G), (G, B), (G, G)}.

   b. $\frac{9}{25}$, which is found by $\frac{3}{5} \times \frac{3}{5}$ as the stages are independent.

   c. $\frac{4}{25}$, which is found by $\frac{2}{5} \times \frac{2}{5}$ as the stages are independent.

   d.   There are also the outcomes where the two colors are different.

21. a.   Since the coins are distinguishable, let the first coordinate of the ordered pair be the outcome for the penny and the second coordinate be that for the nickel. The sample space is then {(H, H), (H, T), (T, H), (T, T)}.

   b. $\frac{1}{2}$.

   c. $\frac{1}{4}$.

   d. $\frac{2}{4}$.

23. a.   Suppose that the data displayed was collected for 100 items. Then of the 10 items that failed test *A*, 3 also failed test *B*.
   Thus the probability that an item that failed *A* also failed *B* is = **0.3**.

   b.   The probability that an item that failed *B* also failed *A* is **3/12 = 0.25**.

   c.   The probability that an item failed *A* or *B* or both is 10/100 + 12/100 − 3/100 = **19/100 = 0.19**.

   d.   Since this is the event that is the complement of *c*, the probability that an item failed neither test is 1 − 19/100 = **81/100**.

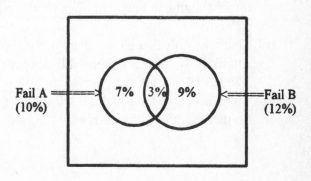

25.

$A = 108$	$B = 108$
$C = 54$	$A = 54$

Given that the chip did not land in B the sample space is $A + A + C$ = 216 sq in. The event space is $A + A = 162$ sq in. So the probability of the chip landing in A given that it did not land in B is **162/216 = 0.75**.

27. Since the probability of a child getting the disease is 0.15 the probability of the complementary event, the child not inheriting the disease is 0.85. So the probability of the first child and the second and the third not inheriting the disease is $(0.85)(0.85)(0.85) = $ **0.614**.

29. There are 216 possible ways the three die can come up. They are related to sums that range from 3, for three 1s showing, to 18, for three 6s showing. Working with the distribution of the 36 possible outcomes on the first two dices sum, we look at a sum of 9 on the first two die. This can happen in 4 different ways. Each of these ways combined with a 6 on the third die results in a sum of 15. There are 2 ways to increase a sum of 10 on the first two dice to a sum of 15 or more. These are the occurrences of a 5 or 6 on the third die. Since there were 3 ways of getting the sum of 10, this gives 6 combinations with an initial sum of 10 to meet or exceed a sum of 15 with three dice. Continuing in a similar way, we find 6 combinations that allow a sum greater than 15 starting with a sum of 11 on the first 2 dice. Finally, there are 4 ways of increasing a sum of 12 on the first 2 dice. This gives a total of 20 ways of getting a sum of 15 or greater on the rolling of 3 dice. Thus, the probability

is $\dfrac{20}{216}$.

31. a. There is 1/3 chance that Bill will pick the top path, which leads to room B. There is 1/3 chance that Bill picks the middle path and then a 1/3 chance that he picks the bottom path at the second fork. Thus, the probability that he ends up in room A this time is $(1/3)(1/2) = 1/6$. There is 1/3 chance that Bill will pick the bottom path and 1/2 chance that he will pick the top path at the second fork. Thus, there is $(1/3)(1/2) = 1/6$ chance that he will end up in room A going this way. So the probability that Bill gets to room A is $1/6 + 1/6 = 2/6 = $ **1/3**.

    b. If Bill selects the middle path the first time, then he has a 1/2 chance of choosing the lower path that leads to room A at the second fork. His probability of ending up in room A is **1/2**.

33. a. Since the first digit is not allowed to be 0 there are 5 possibilities (1, 3, 5, 7, 9) for the first digit and 6 possibilities (0, 1, 3, 5, 7, 9) for the second digit. Since the number must be divisible by 5 the last digit can have only 2 possible values: 0 and 5. So there are $5 \times 6 \times 2 = $ **60** possible 3-digit numbers.

    b. Since the only 3-digit numbers divisible by 25 are _00, _50, and _75, and since the first digit cannot be 0, there are 5 possibilities for the first digit and 3 for the last 2 digits. So there are $5 \times 3 = $ **15** 3-digit numbers.

    c. To be divisible by 100 the last 2 digits must be 00 and there are 5 possibilities for the first digit. So there are **5** 3-digit numbers.

35. Let *S* represent Sue wins a game and *J* represent Jane wins a game. $P(S) = P(J) = 1/2$
    a. The tree diagram is:

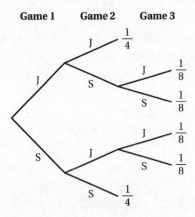

    b. Given that Jane wins the first game, the probability that she will win the second game is 1/2 and thus the probability that she wins the tournament in 2 games given a win in game 1 is **1/2**.
    c. Given that Jane wins game 1, the probability that she will win game 2 is 1/2 and the probability that she will win game 3 is 1/2. So the probability that she wins the tournament in 3 games given a win in game 1 is $(1/2)(1/2) = $ **1/4**.
    d. Given that she loses the first game, the probability that she will win the next two games is $(1/2)(1/2) = $ **1/4**.
    e. The diagram shows that the probability that the tournament goes 3 games is $(1/8) + (1/8) + (1/8) + (1/8) = $ **1/2**.

37. To be successful the missile must penetrate coastal defenses and hit the target. The probability of doing both is the product of the probabilities of doing each independently: $(0.95)(0.8) = $ **0.76**. The admiral should be told that there is a 76% chance that the missile will complete its task successfully.

39. Responses will vary. A possible argument is: Probabilities refer to an infinite number of repetitions of an experiment. If one is tossing a fair coin an infinite number of times one will not obtain all heads or all tails because this would contradict the concept of a fair coin. Thus the probabilities of 0 and 1 should not be associated with the probability of tossing a coin. On the other hand, in the real world, even if the coin is fair it is possible to get as large a run of one face as desired. So probabilities of 0 or 1 should be assigned to such possible runs.

SECTION 9.3

1. There are several approaches to this problem. Theoretically, there are 2 possibilities for the first child, 2 for the second, 2 for the third, and 2 for the fourth: a total of 16 arrangements of gender from BBBB to GGGG. Of these 16, 4 have 3 boys: BBBG, BBGB, BGBB, GBBB. So the probability of three boys and a girl is **4/16, or 1/4**. A second approach is to simulate the birth of 4 children by, say, tossing four coins and letting heads represent the birth of a boy, tails the birth of a girl. One hundred trials of this experiment produced the results:

Number of heads	Number of trials
0	3
1	26
2	45
3	24
4	2

    Thus, the simulated probability of 3 boys out of four children is 0.24.

3. Theoretically, since making and missing free throws are complementary evens, the probability of missing a free throw is 0.75. The probability of missing both the first and the second of a pair of free throws is $(0.75)(0.75) = $ **0.5625**. This can be simulated by selecting 100 groups of 4 random numbers. Let the first 2 of each group of 4 random numbers represent the front end of a pair of free throws and the second 2 numbers represent the second shot. Numbers 00 through 24 represent success in making the free throw, 25 through 99 represents missing. For example 7256 means the player missed both shots and 2181 represents making the first and missing the second.

5. Showing up and not showing up are complementary events. Since the probability of not showing up is 1/6, the probability of showing up is 5/6. So the probability of 7 persons showing up is $(5/6)^7 = $ **0.279**.

7. A possible experiment is flipping a coin 3 times to represent the birth of 3 children: heads representing the birth of a boy, tails the birth of a girl. Repeating the experiment 100 times represents 100 families with 3 children. Sample results are:

Outcome	Frequency	Probability	Outcome	Frequency	Probability
HHH	14	0.14	HTT	8	0.08
HHT	15	0.15	THT	14	0.14
HTH	11	0.11	TTH	13	0.13
THH	12	0.12	TTT	13	0.13

    Because there are 8 equally likely outcomes the theoretical probability of each is **0.125**.

9. Rand produces random numbers between 0 and 1. Six times Rand produces random numbers between 0 and 6 and Int($6 \times$ Rand) produces integers between 0 and 5 inclusive. Thus Int($6 \times$ Rand) + 1 produces the integers 1 to 6 inclusive.

11. To simulate the choice of a locker numbered from 1 to 1000 we must produce random numbers from 1 to 1000. The random number generator of a calculator produced random numbers between 0 and 1. So 1000 × Rand will produce random numbers between 0 and 999.99… Int(1000 × Rand) produces the integers from 0 to 999 inclusive so **Int(1000 × Rand) + 1** will yield the integers between 1 and 1000 inclusive.

13. This situation could be simulated by tossing a die. The numbers 1 and 2 represent toy 1, the numbers 3 and 4 represent toy 2, and the numbers 5 and 6 represent toy 3. Rolling the die once represents a meal. 5 rolls of the die represents one trial of getting five meals. For example, if in 5 rolls the numbers 1, 3, 4, 3, 6, 5 were displayed then toys 1, 2, 2, 2, and 3 would be awarded. Out of 100 trials of 5 meals, all 3 toys were awarded 73 times for a probability of 0.73.

15. To simulate the situation is which the probability of team A winning any one game is 0.7, select random one-digit numbers. A selection of 0 through 6 indicates that A wins the game, 7, 8, 9 that B wins. A series is over when either A or B wins 4 games. Count the number of games in each series. Sample results are:

Games to win series	Frequency
7	8
6	12
5	15
4	15

The average number of games required to win the series is: $[8(7) + 12(6) + 15(5) + 15(4)]/50 = 5$ to the nearest whole game.

17. The selection of all possible males at every choice point is the only way to get the committee. As there are five people on the committee, there is a $\frac{10}{20}$ chance of getting a male on the first selection, $\frac{9}{19}$ chance of getting a remaining male for the second member, $\frac{8}{18}$ of getting a third male, $\frac{7}{17}$ of getting a fourth male, and $\frac{6}{16}$ of getting a fifth male. Thus, the probability of getting a committee consisting of five males is $\frac{10}{20} \times \frac{9}{19} \times \frac{8}{18} \times \frac{7}{17} \times \frac{6}{16} \approx 0.016$.

19. Since the player has a 0.300 probability of getting a hit at each trip to the plate, the probability theoretically theoretically is $(0.300)^4$ of getting a hit on each of four consecutive trips to the plate. This is assuming that the probability of getting a hit at each trip is independent of the associated trips to the plate. This gives a probability of 0.0081 of pulling the 4 consecutive hits off. Using a random generator that gives H a probability of 0.3 and NH a 0.7, repeating the random generation 100 times and recording the results will give a simulated estimate of the probability.

21. The situation may be simulated by drawing with replacement 7 times from the set of numbers 0 through 99. Each draw represents a walk. The 7 draws simulate seven walks. The numbers 0 through 12 represent seeing a snipe, the numbers 13 through 99 not seeing a snipe. The random number generator on a calculator produced the sequence of seven numbers: 17, 99, 82, 88, 31, 30, 08. Thus a

snipe was seen on this sequence of seven walks. Another simulation of seven walks produced: 06, 44, 34, 99, 84, 53, 30. A snipe was also seen on this set of seven walks. Three more simulations of seven walks; 85, 71, 54, 54, 29, 72, 74 (no snipes); 14, 17, 98, 20, 78, 65, 64 (no snipes); 82, 05, 36, 57, 22, 11, 62 (two snipes). So on 5 simulations of seven walks snipes were seen on 3 sets: a probability of 0.6. Theoretically, the probability of never seeing a snipe is $(1 - 0.13)^7 = 0.377$. So the complementary even, seeing a snipe on at least one of seven walks, is 0.623.

23. a.   0.24   b.0.41   c.0.01

25. With $L = 5.5$ cm (a kitchen match) and $D = 8.0$ cm 100 drops of kitchen matches resulted in 41 matches touching or crossing the line. So the experiments produced a probability of 0.41. Since probability $= 2L/\pi D$, we have, as an approximation for π: $2L/Pd = 2(5.5)/(0.41)(8) = 3.35$.

SECTION 9.4

1. The odds in favor of drawing 2 spades on successive draws without replacement can be calculated as the quotient of the probability of drawing two spades and the probability of not drawing 2 spades. The various probabilities may be represented in a tree: So the odds are: [(13/52)(12/51)]/[(13/52)(39/51) + (39/52)(13/51) + (39/52)(38/51)] = **156/2496** = **0.0625**.

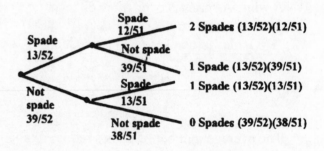

3. The odds against an event A are $\dfrac{1 - P(A)}{P(A)}$. The event A is drawing two hearts on a two card draw.

    $P(A) = (13/52)(12/51) = 1/17$. $1 - P(A) = 16/17$. So the odds are **16 to 1**.

5. Of the 52 possible draws from a deck of cards only 4 result in drawing a king. So $P(K) = 4/52$ and $1 - P(K) = 48/52$. The odds against drawing a king are $(48/52)/(4/2) = 48/4 =$ **12/1**.

7. P(three heads) $= 1/8$. So the odds are $(1 - 1/8)/(1/8) =$ **7 to 1**.

9. 1 to 3

11. 3 to 1

13. The odds, O, in favor of an event A are equal to $P(A) / [1 - P(A)]$. $O = (5/9) / (1 - 5/9) =$ **5/4**.

15. $O = (6/29) / (1 - 6/29) =$ **6/23**.

17. The odd against, X, some event A are $[1 - P(A)] / P(A)$. $X = (1 - 14/37) / 14/37 =$ **23/14**.

19. $X = (1 - 1/671) / 1/671 =$ **670/1**.

21. There are 36 equally likely outcomes for 2 consecutive tosses of 1 die. There are five outcomes that produce a sum of 6: (2, 4), (4, 2), (1, 5), (5, 1), (3, 3). So there are 31 outcomes that will not produces a sum of 6. Therefore the odds in favor of a sum of 6 are **5/31**.

23. Since all outcomes have sums that are greater than 1 the odds in favor of a sum greater than 1 are **36 to 0**.

25. Of the 36 outcomes, 6 give a sum of 7, 30 do not. So the odds are **30 to 6 or 5 to 1**.

27. $\dfrac{5}{6}$, as there are 6 possibilities and 5 are favorable.

29. $\dfrac{5}{12}$, as there are 12 possibilities and 5 are favorable.

31. There are 104 cards with two sevens of spades, 102 non-sevens-of spades. So the odds against drawing a seven of spades are **102 to 2 or 51 to 1**.

33. a. Since there are 2 ways for a green number to show and 36 ways for a non-green number to come up the odds in favor of green are **2/36 = 1/18**.
    b. Of the 38 numbers 0, 00, 1, 2, 3, …, 36 eleven are prime: 2, 3, 5, 7, 11, 13, 17, 19, 23, 29, and 31. So the odds in favor of a prime number winning are **11/27**.
    c. Since 18 numbers are red, the odds in favor of a red number winning are **18/20 = 9/10**.
    d. Because 18 of the numbers are odd, the odds in favor of an odd number winning are **18/20 = 9/10**.

35. We can interpret the 0.4 probability of snow as of 10 weather events that may happen, 4 of them are snow, 6 are not snow. So the odds against it snowing (that is, the odds in favor of no snow) are **6/4 = 3/2**.

37. The expected gain, G, is the difference between premium and expected payout.
    $G = 120 - 0.002(50,000) =$ **\$20**.

39. $P(BC) = 1/26$ and $P(D) = 1/51$. So, assuming the diseases are independent, $P(BC$ and $D)$ $= (1/26)(1/51) = 1/1326$. So the odds that a woman will develop both diseases are: $(1/1326) / [1 - (1/1326)] =$ **1/1325**. The available data about BC on the internet does not suggest that diabetes and BC are related. There is always the possibility that both are related to some common genetic factor.

41. The various expected values, EA, EB, EC, and ED are: $EA = (1/10)(\$2) = 20$ cents; $EB = (2/10)(\$2)$ $= 40$ cents; $EC = (4/10)(\$2) = 80$ cents; $ED = (3/10)(\$2) = 60$ cents. So **A is worst, C is best**.

43. $\dfrac{2 \times 0.10 + 4 \times 0.25}{6} = \$0.20$ as there are six equally likely draws.

45. The odds are (a) 1/12; (b) 1/11 because there are 12 possible outcomes of flipping a coin and rolling a die, one of which is (2, *H*); (c) 16/36; (d) 1/11. From most to least likely the order is: **c, (b and d), a.**

47. The possible equally likely outcomes for flipping 2 coins are: *HH, HT, TH, TT*. The probability of 2 heads is 1/4. The probability of 1 head is 1/2 and the probability of no heads is 1/4.
So *EV* = (1/4)0 + (1/2)10 +(l/4)0 = $5. Since you pay $5 to play, **the game is fair**.

49. Suppose that the individuals are *A, B, C, D, E*, and *F*. Now, *A* had to be born in one of the 12 months. So the probability that *B* was born in a different month is 11/12. And the probability that *C* was born in a month different from both *A* and *B* is 10/12, and so on down to *F* with a probability of 7/12 of being born in a month different from the other 5 individuals. So the probability that *A, B, C, D, E*, and *F* were all born in different months is (11/12)(10/12)(9/12)(8/12)(7/12) = 0.22. The probability of the complementary event, not all were born in different months or at least 2 were born in the same month, is 1 − 0.22 = **0.78**. The same argument applied to nine persons gives a probability that at least 2 were born in the same month as **0.98** and for 12 persons the probability is **0.999995**.

51. A tree diagram for the flips of the three chips is: The diagram shows that of the 8 equally likely outcomes only 2 have all 3 colors different and 6 have 2 colors the same. So, the *EV* for player 1 is (6/8)1 = 3/4 and the expected value for player 2 is (2/8)1 = 1/4. On average player 1 would get the winning 15 points in 20 turns and player 2 would get 15 points in 60 turns. The game is not fair.

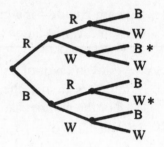

53.

	Odds against	Payout per $2.00 bet
a.	6/1	$14
b.	8/7=1 1/7	$4.29
c.	2/1	$6
d.	19/1	$40

SECTION 9.5

1. a. Discrete, as the number of children is a whole number.
   b. Continuous, as the values, when spread across a population, approximate the real numbers.
   c. Continuous, as the values, could be any real number over a sizable interval.
   d. Discrete, as the number is one of the natural numbers from 2 through 12.

3. As each of the 6 possible outcomes has a $\frac{1}{6}$ probability associated with it, the distribution is discrete and the graph is as shown below:

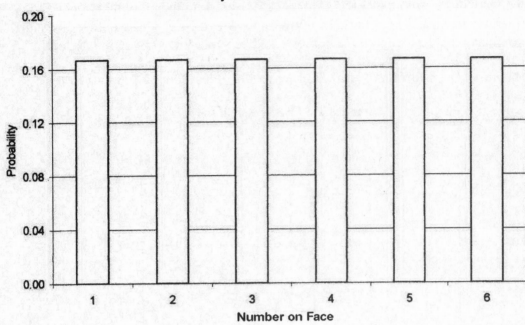

Probability Distribution for Die Roll

5. 0.25

7. a. $z = \dfrac{56 - 50}{4} = 1.5$

    b. $z = \dfrac{19 - 31}{5.5} = -2.182$

    c. $z = \dfrac{57 - 110}{15} = -3.533$

    d. $z = \dfrac{13 - 15}{4.5} = -0.444$

9. The sum of products $1(0.25) + 2(0.30) + \ldots + 8(0.01)$ suggests that the expected value is 2.74 people per household.

11. a. The area of the shaded triangle is 1.
    b. The area of the shaded region to the left of 1.5 is 0.875.
    c. The area of the shaded region to the left of 0.50 is 0.125.

13. The time of 73 seconds has a z-score of -1. Table 9.2 shows 0.3413 of the probability beneath the mean and a z-score of -1. Thus, the probability of being less than 73 seconds is $0.5 - 0.3413$, or 0.1587.

15. a.   15 ounces has a z-score of -1.25. This is associated with the probability of $0.5 - 0.3944$ or 0.1056. This would indicate that approximately 10.56 percent of the bottles could be expected to contain less than 15 ounces.

   b.   17.5 ounces has a z-score of 1.875. This is associated with the probability of $0.5 + 0.4695$ between the mean and 17.5, or 0.306 above the value of 17.5. This would indicate that approximately 3.06 percent of the bottles could be expected to contain more than 17.5 ounces.

17. The probability of a payoff on CC is 0.1111, payoff of at least one cherry is 0.56.

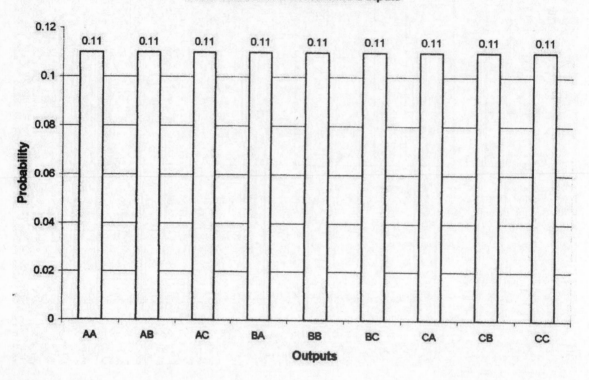

19. Cheryl's z-score was 1, while Becky's z-score was 0.667. Hence Cheryl did better relative to her population than Becky did relative to her population. Assuming the populations are somewhat equivalent, Cheryl did better than Becky.

21. The switchboard will be unable to handle the calls when the load goes over 100 calls. This is associated with a z-score of 1. The area above this point on the score scale is associated with a probability of approximately 0.1587, or 15.87% of the time.

23. Answers will vary.

SECTION 9.6

1. a. If 0 is allowed in the first position then the are 5 possible numbers for the first digit, 5 for the second, 5 for the third, and five for the fourth. There are $5 \times 5 \times 5 \times 5 = $ **625** possible combinations.
    b. Again assuming that 0 may be used in the first position but that repetitions are not allowed, we have 5 possibilities for the first digit, 4 for the second, 3 for the third, and 2 for the fourth. There are $5 \times 4 \times 3 \times 2$, or $_5P_4 = $ **120** possible plates.

3. Since different orders of the same questions are to be considered as different tests, there are $_{14}P_5$ different tests. $_{14}P_5 = 14!/(14-5)! = $ **240,240** different 5-question tests.

5. Since the order of the selections is irrelevant the number of ways is $_{10}C_4$. This is $10!/[(10-4)!4!] = $ **210**.

7. There will be 10 games in the first round, and this will leave 10 teams still in the tournament. In the next round, there will be 5 games, and 5 teams will remain. The next round will have 2 games involving 4 teams (with one team getting a bye). This will leave 3 teams (2 that won in this round and 1 that got a bye). The next round will involve 1 game between a winner and the team that got a bye. The final round will involve 1 game to determine the overall winner. So, there are $10 + 5 + 2 + 1 + 1 = $ **19** games.

9. There are $_4C_2 = 6$ ways of putting two of the coins in the designated pocket. Only one of these, the quarter and dime, give more than 30 cents. So the probability is **1/6**.

11. The coins may be selected individually, in pairs, in threes, in fours, and finally in 1 group of 5. The order of selection is not important and no grouping has the same value as any other grouping. So the total number of combinations is:
    $_5C_1 + _5C_2 + _5C_3 + _5C_4 + _5C_5 = 5 + 10 + 10 + 5 + 1 = $ **31**.

13. Ignoring order, the number of ways of selecting 9 persons from a pool of 20 persons is $_{20}C_9$ which is $20!/(20-9)!9! = $ **167,960**.

15. Since order is not important we have: $_{52}C_5 = 52!/(52-5)!5! = $ **2,598,960** different 5-card hands.

17. There are 6: RS, RT, SR, ST, TS, and TR

19 11!, as there are 11 people.

21. Represent the license plate number $D1$ $D2$ $D3$ $D4$. $D1$ must come from the set $\{1, 2, 3, 4, 5\}$. Thus there are 5 possibilities for $D1$. Now, for each $D1$ there are 5 possibilities for $D2$ from the set $\{0, 1, 2, 3, 4, 5\}$. There are 5 possibilities, not 6, because consecutive digits cannot repeat because license plates must be unique. $D4$ has 3 possibilities, the set $\{0, 2, 4\}$, because the license number must be even. So that leaves 4 possibilities for $D3$ from the set $\{0.1, 2, 3, 4, 5\}$. There are only 4 possibilities because $D3$ cannot repeat on the left or the right. The total is $5 \times 5 \times 4 \times 3 = $ **300**. The same argument holds for odd numbers except that $D4$ must come from $\{1, 3, 5\}$. So there are also **300** different odd license plates.

23. Suppose the offices are P, V, S, and T. P can be filled in 6 ways, V in 5 ways, S in 4, and T in 3. So the total is $6 \times 5 \times 4 \times 3 = \textbf{360}$, or $_6P_4$.

25. Because order is not important the total number of different committees is $_{57}C_8 \times {_{43}C_4} = \textbf{2.039} \times \textbf{10}^{14}$.

27. There are 6 possible colors to face up. For each of these 6 there are 4 colors to face forward.

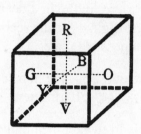

Up	Forward
R	OYGB
V	OYGB
0	BRYV
G	BRYV
Y	ORGV
B	ORGV

So there are **24** orientations.

29. $_nC_r = n!/[(n-r)!r!]$ and $_nC_{n-r} = n!/[n-(n-r)]!(n-r)! = n!/r!(n-r)!$. So $_nC_r = nC_{n-r}$.

31. The combinations of 60 people taken 12 at a time. This gives $\dfrac{60!}{12!48!}$ which is approximately 1,399,400,000,000 ways.

33. 10 ways. Easiest seen with a tree diagram.

35.

n	$(1+1)^n$
1	2
2	4
3	8
4	16

$(1+1)^n$ is a geometric progression with $r = 2$.

37. Responses will vary. Suppose the 8 condiments were add/no-add offer. Then there are $2 \times 2 \times 2 \times 2 \times 2 \times 2 \times 2 \times 2 = 256$ combinations. If, in addition, there are 3 types of buns and 3 kinds of cheese, then we have $2 \times 2 \times 2 \times 2 \times 2 \times 2 \times 2 \times 3 \times 3 = 1152$ varieties. So the company is telling the truth.

39. There are $3! = 3 \times 2 \times 1 = 6$ ways to seat 4 people at a circular table. There are $4!$ ways to seat them, but we need to divide by 4 in order to eliminate rotations that produce the same seating arrangement.

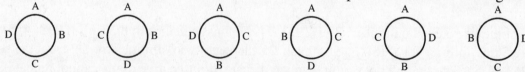

41. There are $_{15}P_4 = 32,760$ ordered combinations of individuals who could be interviewed. There are $_{15}C_4 = 1365$ unordered combinations of individuals who could be interviewed. It is not reasonable for an interview committee to take the time to generate these lists of combinations and then choose one combination of individuals to interview. If they could list 1 permutation every 15 seconds, it

would take them over 5 days just to generate the list of candidates. It would take over 5 hours to list all of the combinations.

43. There are 120 wires on either side of the cut. So if you pick a wire on one side, you could connect it to any one of 120 wires on the other side. The next wire can be connected to any of the 119 remaining wires. The third wire can be connected to any of the remaining 118 wires, etc. So there are $120! \approx 6.69 \times 10^{198}$ ways the cable could be spliced. Using the same reasoning, if the wires were bundled into 10 groups of 12, there would be 10 choices for a connection for the first bundle, 9 choices for a connection for the second bundle, etc. Thus, there are $10! \approx 3.6 \times 106$ possible ways to connect the bundles. Then, within each bundle there are 12 wires to be connected, and these can be connected in $12! \approx 4.8 \times 10^8$ ways. So, there are $(3.6 \times 10^5) \times (4.8 \times 10^8) = 1.7 \times 10^{15}$ ways the wires can be connected in bundles. This is far fewer options than when the wires are separate.

CHAPTER 9 REVIEW EXERCISES

1. a. The sample space for tossing three coins is: {HHH, HHT, HTH, THH, TTH, THT, HTT, TTT}.
    b. The subset with T in the second position is {HTH, TTH, HTT, TTT}.
    c. The event H on the third toss is {HHH, HTH, THH, TTH}.
    d. The event of H on the second toss and T on the third toss is {HHT, THT}.

3. a. $P(JD \text{ and } E) = (1/52)(3/6) = \mathbf{1/104}$.
    b. $P(R \text{ and } n \geq 3) = (26/52)(4/6) = \mathbf{1/3}$.
    c. $P(F \text{ and } 5) = (16/52)(1/6) = \mathbf{2/39}$.

5. a. The uniform sample space for tossing 2 dice is {11, 12, 13, 14, 15, 16, 21, 22, 23, 24, 25, 26, 31, 32, 33, 34, 35, 36, 41, 42, 43, 44, 45, 46, 51, 52, 53, 54, 55, 56, 61, 62, 63, 64, 65, 66}. Of these the sum 7 can be obtained in 6 ways, the sum 8 in 5 ways, the sum 9 in 4 ways, the sum 10 in 3 ways, the sum 11 in 2 ways, and the sum 12 in 1 way. Thus the event space for a sum of 7 or greater has 21 elements and a probability of **21/36**.
    b. Two of the 36 elements of the sample space show one 3 and one five. So the probability is **2/36**.
    c. The first die can be 1, 3, or 5 and for each of these the second can be 1, 3, or 5. So there are 9 outcomes with both numbers odd. So the probability is **9/36**.
    d. Of the 36 outcomes, 9 are even/even, 9 are odd/odd, 9 are even/odd, and 9 are odd/even. So 27 outcomes have at least 1 die even. Additionally, there are 3 odd/odd combinations with a sum of 6. Thus the event has 30 of the 36 outcomes and a probability of **30/36**.

7. Since this situation deals with two draws without replacement, the sample space, as opposed to problem #2 in this section, will change after the first draw.
    a. The probability of 2 red marbles is **0** because if a red marble is drawn first it is not replaced. So the sample space for the second draw does not contain any red marbles. Thus the probability of a red on the second draw is 0.
    b. Since $P(R \text{ on } 1^{st})$ is 1/4 and $P(G \text{ on } 2^{nd} \text{ given } R \text{ on first})$ is 1/3, the probability of first drawing red and then drawing green is $(1/4)(1/3) = \mathbf{1/12}$. An alternative approach is to examine the 12 equally likely outcomes; for each possible 1st draw there are 3 equally likely second draws, a total of 12. One of these 12 is $R - G$.

    c.   The sample space of 12 equally likely outcomes is *R-B, R-G, R-Y, B-R, B-G, B-Y, G-R, G-B, G-Y, Y-R, Y-G, Y-B*. Note that 2 of the 12 have at least one *R* and one *G*. Thus the probability is **2/12 = 1/6**.

    d.   Six of the 12 equally likely outcomes have no yellow (*Y*). Thus the probability of no *Y* is **6/12 = 1/2**.

9. Expected value = sum of the (probability-times-outcome) products.
    So $EV = 0.15(0) + 0.35(-10,000) + 0.2(-20,000) + 0.3(55,000)$. So the *EV* is **\$9,000**.

11. a. Examination of the demographics shows that the populations of the states are far from equal. Comparisons of the populations suggest how likely it is for a person to live in a particular state. So one cannot consider a listing of the states to be a uniformly distributed sample space.

    b.   These events are not mutually exclusive. The same person can attend both a 2-year and a 4-year institution of higher learning.

    c.   Because missing a free throw and making a free throw are complementary events, the sum of the probabilities must be 1.

    d.   One might think that they must be the ill person, but this WRONG! The quoting of statistics like this applies in general to random groups of four selected from the population or, said differently, on average we expect this to be true for random groups of four people. It does not mean in every group of four people.

13. The teacher has 15 choices for seating a student at the first desk, 14 for the second desk, and so on down to 1 student at the last desk. So there are **15!** different arrangements possible: about $\mathbf{1.3 \times 10^{12}}$.

15. Mutually exclusive events are those that have no common outcomes in the two event spaces. As an example, the events of an even number on a roll of a die and an odd number on the same roll of the die are mutually exclusive. The events of rolling an even number and rolling a prime number are not mutually exclusive because one even number, 2, is also a prime. Independent events are events in which the outcome of one does not affect the outcome of the other. For example, if the same die is rolled twice the outcome of the second roll is not affected by the outcome of the first roll. The probability of obtaining, say, a six on the second roll is 1/6 no matter what happened on the first roll. The probability of the second roll producing an even sum obviously is dependent upon the outcome of the first roll of the die.

17. The prime factorization of 105 is $3 \times 5 \times 7$. One of these numbers is the number of meats, another the number of toppings (onions, peppers, olives, etc.), and the third the number of cheeses. It is reasonable to assume 5 meats and 7 toppings. So that leaves **3** for the number of cheeses.

19. Assume that the kernel of corn falls either into one of the silos or into the inner area bounded by the silos. Each silo has an area of $(3.14)(20)^2$, 1256 sq ft. So the four together have an area 5024 sq ft. The square formed by joining the centers of the silos has an area of 40 ft × 40 ft = 1600 sq ft. Each of the shaded portions is one quarter of a silo circle and, all together, make up one circle with an area 1256 sq ft. So the event space, the area inside the small square but not in the silos has an area of 1600 − 1256 = 344 sq ft. Thus the probability of landing in this area is **344/(5024 + 344) = 0.064**.

21. A tree diagram begins 3 blank branches. For each of these blank branches there are 5 position branches. For each of the position branches there are 3 branches; either do not remove metal at that position or remove it at one of two different depths. So there are (3)(5)(3), **45**, different keys.

23. The area of the bull's-eye, the event space, is 4π. The area of the first dark ring is $(36 − 16)\pi$, the area of the second dark ring is $(100 − 64)$pi. So the entire shaded area, the sample space, has an area of $(4 + 20 + 36)\pi$. Thus the probability of a bull's-eye given landing in the shaded area is $4\pi/(60\pi) = \mathbf{4/60 = 1/15}$.

SECTION 10.1

1. a. point   b. line segment   c. ray   d. parallel lines   e. line   f. octagon   g. angle   h. circle

3.

5.

7.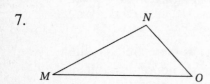

9. Answers will vary: $\overline{GH}, \overline{GK}, \overline{GD}$

11. *EL* and *KM*

13. *DFG* (right: contains a right angle), *HIJ* (acute: all angles acute), *DFH* (scalene: no two sides equal).

15. *GHJLK*.

17. The sum of the two angles is a straight angle.

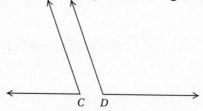

19. Angles 1 and 2 are formed by intersecting lines, have a common vertex, and have no common side.

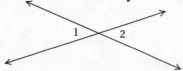

21. **False.** Since concurrent lines contain a common point and since parallel lines have no common points, concurrent lines cannot be parallel.

23. False. Skew lines do not intersect. Perpendicular lines intersect to form right angles.

25. **False.** Two radii that do not form a straight do not make a diameter of a circle.

27. **False.** Since sides opposite equal angles in a triangle are congruent and since all the angles of an equiangular triangle are congruent, all the sides are also congruent.

29. **False.** Since the sum of all three angles of a triangle is 180 degrees, no two could have a sum of 180 degrees.

31. Since the lines, r, s, and t are concurrent figures, they intersect at a point and thus are not parallel.

33. **No.** By definition, skew lines cannot be contained in the same plane. Were they to intersect, they would determine, and thus be contained in, a single plane.

35. Points A, B, and C must be collinear with B and C on the same side of A.

37. The Venn diagram shows that all squares are rhombi and also that all rhombi are kites. However, some kites are not rhombi and some rhombi are not squares.

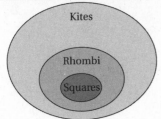

39.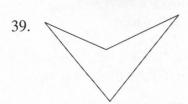

41. A linear pair of angles has a sum of 180 degrees. Since each obtuse angle has a measure greater than 90 degrees, the sum of two obtuse angles is greater than 180 degrees. **So two obtuse angles cannot be a linear pair**. Since an acute angle has a measure less than 90 degrees, the sum of two acute angles is less than 180 degrees. **So two acute angles cannot be a linear pair**.

43. Responses will vary.

45. The ratio 5:3 is 1.6666..:1. The ratio 8:5 is 1.6:1. The Golden Ratio is approximately 1.618. The **5 by 8** card has the closer ratio.

47. a. The rectangle contains 5 rows of 8 squares each for a total of 40 squares.
    b. A(rectangle $ABCD$) $= l \times w = 8 \times 5 = 40$.

49. a. Triangles ADF and EBC together contain 8 unit squares. $AECF$ is a square containing 16 units, so the total area of the parallelogram is 24 square units.
    b. A(parallelogram $ABCE$) $= b \times h = 6 \times 4 = 24$ units.

51. Designs will vary. Here is a sample.

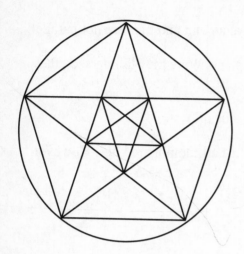

53. The Fibonacci sequence begins 1, 1. Succeeding terms are formed by adding the two preceding terms. The first 15 terms are: **1, 1, 2, 3, 5, 8, 13, 21, 34, 55, 89, 144, 233, 377, 610**.

55. Some terms of the Fibonacci sequence are 1, 1, 2, 3, 5, 8, 13, 21. Now, $1^2 = 1$, $1 \times 2 = 2$; $3^2 = 9$ and $2 \times 5 = 10$; $5^2 = 25$ and $3 \times 8 = 24$. $8^2 = 64$ and $5 \times 13 = 65$. Finally, $13^2 = 169$ and $8 \times 21 = 168$.

Square of a term	1	4	9	25	64	169
product of preceding and succeeding terms	2	3	10	24	65	168

It appears that the square of a term of the Fibonacci sequence plus or minus 1 is equal to the product of the preceding and succeeding terms.

57. Responses will vary. One possible sequence is:

x	1.5	1.25	1.75	1.7	1.65	1.6	1.61
y	0.5	0.25	0.75	0.7	0.65	0.6	0.61
$x - y$	1	1	1	1	1	1	1
xy	0.75	0.31	1.31	1.19	1.07	0.96	0.98

It appears that x is approaching the golden ratio and that y is its reciprocal.

59. **Never**. The sum of the angles about a point is 360°. Two intersecting lines form 4 angles. If they are congruent, then each must be 90°, a right angle.

61. **Always**. The exterior sides of linear angles form a straight angle. If the sum of 2 angles is a straight angle, then the angles are supplementary.

63. **Sometimes**. Linear angles are supplementary angles with a common vertex. But any pair of angles with a sum of 180° are supplementary, part of the same figure or completely distinct.

65. **Sometimes**. If the rhombus has a right angle, then it is a square.

67. **Never.** If a kite had a pair of parallel sides as does a trapezoid, then the kite would have 2 pairs of parallel sides making it a parallelogram. But no trapezoid is a parallelogram.

69. **Sometimes.** A parallelogram that is both a rectangle and a rhombus is a square. But not all parallelograms are squares.

71. **Never.** Because a parallelogram has 2 pairs of parallel sides, no parallelogram is a trapezoid.

73.

m∠1 = 70°. m∠1 + m∠2 = 90°. So m∠2 = 20°.
m∠2 + m∠3 = 90°. So m∠3 = 70°. m∠3 + m∠4 = 90°.
So m∠4 = 20°.
To check: m∠1 + m∠2 + m∠3 + m∠4 = 180°.
70° + 20° + 70° + 20° = 180°.

75. The processes will depend upon the GES used. In general, (**a**) Congruent figures may be obtained by using a form of the COPY command. (**b**) The COPY command applies to all types of figures, angles and segments. (**c**) Most GES have a BISECT command that will determine midpoints and angle bisectors. (**d**) Same as (**c**). (**e**) Once a line and point have been drawn, GES has a construct PERPENDICULAR command (**f**) Same as (**e**).

77. a.   Not possible for 4 lines to intersect in exactly 2 points.

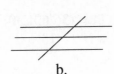 b.   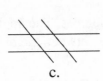 c.   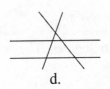 d.   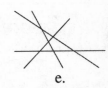 e.

79. Responses will vary

81. Answers will vary. Possibilities include.

 a   b   c   d   e

83.

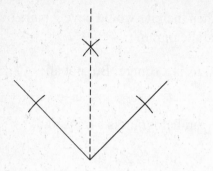

85.

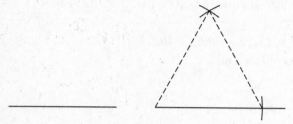

87.

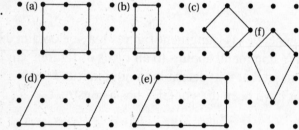

89. The angle $3O7 = (4/12)360 = 120$ degrees. The angle $3OA = (36/60)(30) = 18$ degrees. The angle $BO7 \ (1/5)30 = 6$ degrees. The angle between the hands is $120 - 18 + 6 = 108$ degrees.

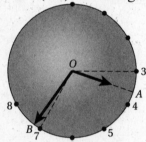

91.

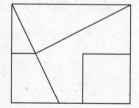

93.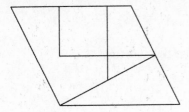

95.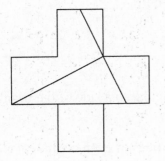

97. 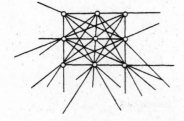   **20 lines**

99. The first office must be connected to seven others: 7 cables. The second office must now be connected to six more offices: six more cables. The third must be connected to 5 more offices: five more cables. The seventh will have to be connected to only one more office and the last office, will already have been connected to all. So the total number of cables is
$7 + 6 + 5 + 4 + 3 + 2 + 1 = $ **28 cables**.

101. This is a student activity (Appendix B).

103. a.   This is an activity. (b) Since it is a surface with exactly one edge and one side, when cut down the middle it remains in one piece, as a loop with two twists in it.

105. a. $A3 = A1 + A2$
   b.   After entering "$= A1 + A2$" in cell 3, highlight cells $A3$ through $A15$ and use "fill down" to place the correct formula in each cell $A4 - A15$.

SECTION 10.2

1. A segment connecting a vertex with the midpoint of the side opposite that vertex.

3. The point of intersection of the three altitudes of the triangle.

5. The point of intersection of the bisectors of the three angles of the triangles.

7. 15 units.

9. $DC = 12$, $CM = 6$.

11. Construct the incenter, and use the distance from the incenter perpendicular to a side of the triangle as radius to draw a circle.

13. Add another trail from the rest area to the ranger station making all vertices of even degree.

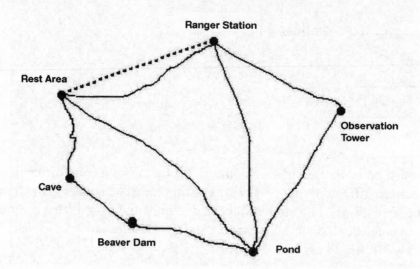

15. **Yes**. All the vertices are of even degree.

17. Student activity.

19. Student activity.

21. These three points are concurrent.

23. In an equilateral triangle, the centroid, orthocenter, incenter, and circumcenter are all at the center point of the triangle.

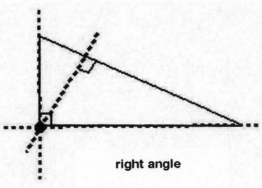

right angle

25. When the triangle is an isosceles right triangle.

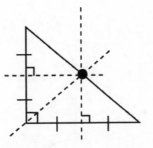

isosceles right triangle

27. The ortho center and circumcenter.

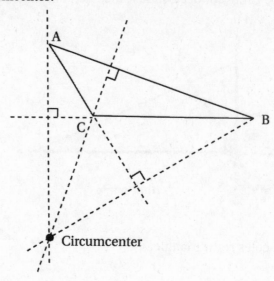

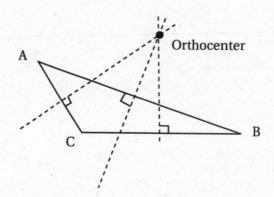

29. Responses will vary. Because all the vertices are even, there is a circuit around the figure. Three such paths are: *ABCDEFBDFACEA*, *AFEDCBFDBAECA*, and *AECAFDBFEDCBA*.

31. They are all on the same circle.

33. The generalization is true.

35. The bridge between B and C is removed, and placed between B and D. Now the network has exactly two odd vertices, and is traverable, starting at A and ending at B.

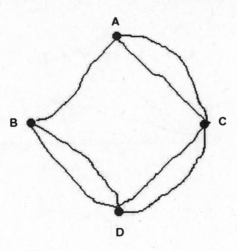

37. vertex = state
    edge = crossing border

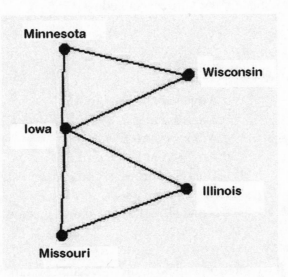

Since the network has all even vertices, it is traversable type 1, and you can start at any state, cross each border exactly once, and return to where you started.

39.

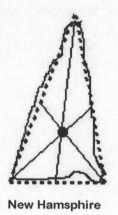

New Hamsphire

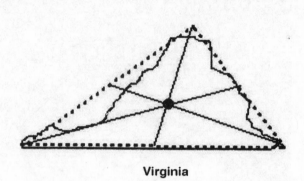

Virginia

41. Details will vary. Construct the nine point circle, and the circumscribed circle for a triangle. Then measure to compare their radii.

43. Answers may vary. For example, a network was used in Exercise 38 to connect a map of five states to a mathematical network made up of vertices and edges.

SECTION 10.3

1.

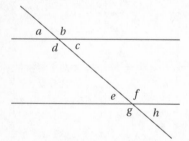

  a. **Angles d and e, c and f.**
  b. **Angles a and e, c and h, b and f, d and g.**
  c. **Angles a and h, b and g.**
  d. **Angles d and e, c and f.**
  e. **Corresponding a and e, alternate interior c and e, alternate exterior a and h.**
  f. **Same side exterior a and g, same side interior c and f.**

3. $2x - 7 + 3x + 2 = 180$, $5x - 5 = 180$, $5x = 185$, $x = 37$. So $m\angle 4 = 2(37) - 7 = 67°$, and $m\angle 5 = 3(37) + 2 = 113°$. Two interior angles on the same side of a transversal cutting two parallel lines are supplementary.

5. $3x - 4 + 9 + 16 = 180$. $12x + 12 = 180$, $12x = 168$, $x = 14$. $m\angle 7 = 3(14) - 4 = 38°$. $m\angle 2 = 9(14) + 16 = 142°$. Same side exterior angles are supplementary.

7. $\angle$s 13, 12, 9, 1, 4, 5, 8.
  a. Vertical angles are congruent.
  b. Alternate interior angles are congruent.
  c. Corresponding angles are congruent.

9. 15, 7, 2
  a. Corresponding angles are congruent.
  b. Alternate exterior angles are congruent.

11. 2, 7, 10, 15.
   a. Same side exterior angles are supplementary.
   b. Alternate exterior angles are congruent.
   c. Two adjacent angles whose noncommon sides lie on a straight line are supplementary.

13. Angles 3 and 52 deg are supplementary so angle 3 = 128 deg. The sum of the angles of a triangle is 180 deg so angle 1 + angle 2 + 52 deg = 180. Thus angle 1 + angle 2 = 128 deg.

15. Let S = the sum of the interior angles of a polygon with n sides. $S = (n - 2)180$.

17. Interior angle, I, = $180(n - 2)/n$; exterior angle, E, = $180 - I$; central angle, $C = 360/n$.
   $I = 180(5 - 2)/5 = 108$, $E = 180 - 108 = 72$, $C = 360/5 = 72$.

19. Interior angle, I, = $180(n - 2)/n$; exterior angle, E, = $180 - I$; central angle, $C = 360/n$.
   $I = 180(10 - 2)/10 = 144$, $E = 180 - 144 = 36$, $C = 360/10 = 36$.

21. Interior angle, $I = 180(n - 2)/n$; exterior angle, E, = $180 - I$; central angle, $C = 360/n$.
   $I = 180(9 - 2)/9 = 140$, $E = 180 - 140 = 40$, $C = 360/9 = 40$.

23. The measure of the interior angles increases getting closer to 180 deg. The measure of the central angles decreases getting closer to 0 deg.

25. The sum of the interior angles of a triangle is 180 deg. So $37 + 59 + c = 180$. $c = $ **84 deg**.

27. Since the shape is a pentagon, the sum of the interior angles is $3(180) = 540$ deg. The two non-right angles are congruent. So $540 = 3(90) + 2x$. $x = $ **135 deg**.

29. The measure of angle P is one half the difference of the intercepted arcs: $\frac{1}{2}(90 - 20) = $ **35 deg**.

31. Angle P, formed by intersecting chords, is equal to one half the sum of the intercepted arcs:
   $\frac{1}{2}(42 + 12) = $ **27 deg**.

33. Angle P, formed by a tangent and a secant, is equal to one half the difference of the intercepted arcs:
   $\frac{1}{2}(144 - 62) = $ **41 deg**.

35. Since the pentagon is regular, the sides (chords of the circle) are congruent and the arcs AB, BC, CD, DE, EA all measure $360/5 = 72$ deg. Angle DAC is an inscribed angle intersecting a 72 degree arc and thus measures **36 deg**.

37. A stop sign is a regular octagon. The interior angles are all congruent and equal to $(8 - 2)(180)/8 = $ **135 deg**.

39. a. The sum of the three angles of any triangle is exactly 180 degrees. By definition the measure of an obtuse angle is greater than 90 degrees and the sum of 2 obtuse angles is greater than 180 degrees. Thus no 2 obtuse angles may be in the same triangle.
   b. The sum of any obtuse angle and a right angle is greater than 180°. But the sum of the angles of a triangle is exactly 180°. So no triangle may contain both a right and an obtuse angle.

41. A triangle has 6 exterior angles. Usually the exterior angles are considered in groups of 3 as shown in the figures.

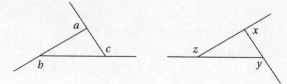

43. The sum of the measures of angles 1, 2, and 10 and also of angles 5, 6, and 7 is 180 degrees because they are sets of interior angles of triangles. The sum of the measures of angles 3, 4, 8, and 9 is 360 degrees because they are a set of interior angles of a quadrilateral. Thus the sum of the measures of angles 1 through 10 is 720 degrees. This sum can be written as $m\angle 1 + (m\angle 2 + m\angle 3) + (m\angle 4 + m\angle 5) + m\angle 6 + (m\angle 7 + m\angle 8) + (m\angle 9 + m\angle 10) = 720°$. Written this way, the sum, 720°, of the 6 terms is the sum of the interior angles of the hexagon.

45. Each interior angle of a regular pentagon measures $(5 - 2)180/5 = 108°$ and each interior angle of a regular decagon measures $(10 - 2)(180)/10 = 144°$. The sum of two interior angles of regular pentagons and 1 interior angle of a regular decagon is $(108 + 108 + 144) = 2(108) + 144 = 360°$. So the figures will fit around a point with no gaps or overlaps.

47. Let n represent the number of sides. Then an exterior angle measures $360/n$ degrees and an interior angle measures $(n - 2)(180)/n$. Since the interior angle is to be 8 times the exterior angle we have $8(360/n) = 180 - (360/n)$. So $n = 9(360)/180 = \mathbf{18}$. Checking, an exterior angle is $360/18 = 20$ degrees and an interior angle is $(16/18)(180) = 160$ degrees. And 160 is 8 times 20.

49. It is a square.

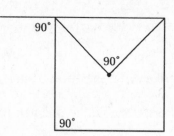

51. a. They meet at an exterior point.
    b. Yes, the discovery holds. The bisectors of two exterior angles of a triangle and of the third exterior angle are concurrent at a point outside the triangle.
    c. The point of concurrency in (b) is the excenter of the triangle. The three excenters are centers of three circles termed excircles.

53. If the quadrilateral is a rhombus, a kite, or a square then the diagonals are always perpendicular, but for a rectangle that is not a square, the diagonals are not perpendicular. So the diagonals of a convex quadrilateral are **sometimes** perpendicular.

55. GES gives:

$\angle G$	$\angle H$	$\angle I$	$\angle H + \angle I$
90	20	70	90
90	50	40	90
90	75	15	90

It appears that angles H and I are complementary.

57. It appears that the sum of the measures of the opposite angles is 180°. The use of GES supports this conjecture.

59. Lines p and q are perpendicular to line m, so corresponding right angles are formed. Since right angles are congruent, we conclude that since p and q are cut by a transversal m so that corresponding angles are congruent, we know from the theorem that p is parallel to q.

61. 36°, 72°, 108°, 144°. EF ∥ GH, since $\angle$E and $\angle$H, as well as $\angle$F and $\angle$G are supplementary, and when two interior angles on the same side of the transversal are supplementary, the lines are parallel.

63. The sum of the measures of the exterior angles of any convex polygon is 360 degrees. So the sum of the interiors, in this case, must be 720 degrees. Now, the sum of the interior angles for this regular polygon of n sides is $(n-2)180$ with n representing the number of sides. So $(n-2)(180) = 720$; $n = \mathbf{6\ sides}$.

65. Place the square as shown and mark points A and B and points C and D. Draw AB and CD. Because X and Y are right angles, AB and CD are diameters. Diameters intersect at the center of a circle.

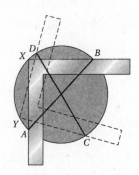

67. Arguments will vary. One possibility is: All of the exterior angles together make up a rotation of 360 degrees. As the number of sides increases the rotation is made in an increasing number of smaller steps. On the other hand, as the number of sides increases, the number of interior/exterior pairs increases. The sum of each pair is 180 degrees so the sum of all the pairs increases with an increase in the number of sides. But since the sum of the exteriors is a constant, 360 degrees, the sum of the interiors increases with increasing number of sides to provide the increasing total of the pairs.

69. 7.5°: 500 = 360° : C. So C = 24,000. The circumference of the Earth is given as 24,902 miles. Both Erathosthenes' linear and angular measure lacked the precision available today.

71. For each of the n sides there is a triangle with an angle sum of 180 degrees. So the sum of all the angles is 180 n. Subtracting the sum of all the angles at the point inside the polygon, 360 degrees, we are left with $180 n - 360 = 180 n - 2(180) = (n-2)180$. This sum is the sum of all the remaining angles of the n triangles which make up the interior angles of the polygon.

SECTION 10.4

1. Triangles ABC and DEF are congruent by SAS.

3. Triangles SRO and BAT are congruent by AAS.

5. The triangles are not necessarily congruent.

7. 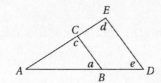 Segments *CB* and *ED* are parallel so angles *c* and *d* are congruent and angles *a* and *b* are congruent. Angle *A* is congruent to itself. So triangles *ACB* and *AED* have angles respectively congruent but the triangles are not congruent.

9. Triangles BDM and FDM are congruent by *SAS*.

11. Triangles *BAJ* and *FGH* are congruent by *AAS*.

13. Triangles *KBM* and *LFM* are congruent by *HL* or *SSS* or *SAS*.

15. No. The sides within an equilateral triangle are congruent, but the sides between equilateral triangles are not necessarily congruent, so the triangles are not necessarily congruent.

17. Not similar.

19. Not similar.

21. $\triangle$ABC $\sim \triangle$DEF because of the SSS Similarity Theorem.

23. $\angle$B and $\angle$E are still congruent in the figure below, and BC and EF are in the same ratio. But the length of ED has been extended so that the corresponding sides of the two triangles are not proportional and hence the triangles are not similar.

25. $d^2 = 5^2 + 5^2 = 50$
    $d = \sqrt{50} = 5\sqrt{2}$

27. In order to find *d*, we must first find the length of the third leg of the triangle (half of the length of the diagonal of the base). The base is a 6×6 square, so the length of the diagonal is $\sqrt{6^2 + 6^2} = \sqrt{72} = 6\sqrt{2}$. So the leg of the triangle is half of that length or $3\sqrt{2}$. Applying the Pythagorean Theorem to determine the length of *d* we get $6^2 = d^2 + (3\sqrt{2})^2$; $36 = d^2 + 18$; $18 = d^2$; $d = 3\sqrt{2}$.

29. Responses will vary. Possibilities include: suppose *a*, *b*, and *c* are the sides of a right triangle with *c* the side opposite the right angle. From an algebraic point of view:. Then $c^2 = a^2 + b^2$. From a geometric point of view, the area of the square with side *c* is equal to the sum of the areas of the squares with sides *a* and *b*.

31. Since the lot is square, by the properties of special right triangles, the length of the diagonal is $\sqrt{2}$ times the length of one side. So the length of the diagonal is $\sqrt{2} \times 240 \approx 1.414 \times 240 \approx 339.36$ feet.

33. $d^2 = 18.5^2 + 26^2$, $d^2 = 342.25 + 676 = 1018.25$. So $d = \sqrt{1018.25} = 31.91$, and the TV is 32″.

35. Using the Pythagorean theorem we have the height, h, is $\sqrt{[6^2 - 2^2]} = 4\sqrt{2} \approx$ **5.66 feet**.

37. The diagonal of the doorway is
    $\sqrt{(6.5^2 + 3^2)} \cong 7.16\,\text{ft}$. So the 7 ft square piece of
    plywood **can be carried** through the doorway.

39. The perimeter, p, is $\sqrt{(1^2 + 3^2)} + \sqrt{(1^2 + 6^2)} + \sqrt{(1^2 + 3^2)} + \sqrt{(1^2 + 4^2)} \approx 16.53$ units. Since 1 unit
    represents 25 ft, the hedge length should be $25 \times 16.53 \approx$ **413.25 ft**.

41. Since the segment joining the midpoints of two sides of a triangle is half the third side, the ship is
    twice the length of the barge: **850 feet**.

43. She measured the correct distances on the ground. Then she used similar triangles and solved the
    proportion, $h/5 = 72/8$, or $h = 45$. The flagpole is 45 ft. tall. Since $\angle i \cong \angle r$, the right triangles are
    similar because of the Right Triangle Similarity Theorem. Because they are similar, their sides are
    proportional.

45. $h = 2 \times 8 =$ **16**, $s = 8\sqrt{3}$.

47. The leg of the right triangle, which is the side opposite the 30 deg angle, is $5\sqrt{2} / \sqrt{2} = 5$. So
    $x = 2 \times 5 =$ **10** and $y = 5\sqrt{3}$.

49. $\dfrac{2\sqrt{3}}{3}$ See # 33.

51. The legs of a right triangle include the right angle and all right angles are congruent so the triangles
    are congruent by **SAS**.

53. The right angles are also congruent. So, the right angle, an acute angle, and the hypotenuse are
    congruent between the two triangles. Thus, by **AAS** (because the hypotenuse is not between the
    2 congruent angles), the triangles are congruent.

55. a. **Never**. There can be no right angle.
    b. **Sometimes**. A right triangle may have congruent legs and be isosceles.
    c. **Sometimes**. A right triangle may have no sides congruent.
    d. **Always**. The Pythagorean Theorem is specific to right triangles.

57. *FG* may be considered to be the hypotenuse of a right triangle with
    legs *FH* and *HG*. Now, the length of *FH* is the absolute value of the
    difference between the *x* coordinates of points *F* and *H* and the
    length of *GH* is the absolute value of the difference between the *y*
    coordinates of points *G* and *H*. Because the differences will be
    squared in the computation of the length of *FG*, the absolute value
    restriction may be dropped and the differences determined in either

    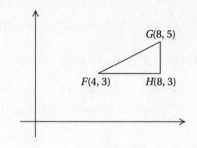

    order. So $FG = \sqrt{[(8-4)^2 + (5-3)^2]} = 2\sqrt{5} \approx \mathbf{4.47}$.

59. Responses will vary.
    All the triangles can be
    formed on a 3 by 3
    Geoboard.

    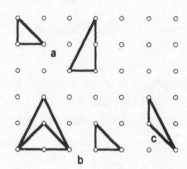

61. Because the radii meet at a right angle, we know that the arc formed by the wood is 1/4 of the
    circumference of the circle. Thus, the circumference is 64 feet, and the radii must be ~10.19 feet
    (solving $C = 2\pi r$). Since the right triangle is isosceles, we know that the length of the hypotenuse is
    $\sqrt{2}$ times the length of the leg or (10.19)($\sqrt{2}$) or approximately 14.4 feet.

63. GES suggests the figure is an
    equilateral triangle.

65. GES shows that *PQ* is one half of segment *CB* and also that the four triangles are congruent to each
    other and to the original triangle.

67. Answers will vary. A possibility is: Right triangles contain right angles; all right angles are
    congruent; triangles are *AAS* congruent.

69. The angle of the points is found by subtracting the acute angles of the right triangles (sum = 90 deg)
    from the interior angles of the regular heptagon. But all the interior angles are congruent and
    measure (7 – 2)180/7 ≈ 128.6 deg. All the right triangles are congruent (*HL*). So all the point angles
    are formed by subtracting the same two values from 128.6.

71. The 2 large squares both have sides of length *a* + *b* and so have the same area. Two of the triangles
    cover one of the rectangles so if all 4 triangles are removed from one large square and both
    rectangles are removed from the other large square the remaining areas, a square of side *c*, and two
    areas, squares of sides *a* and *b*, are equal.

73.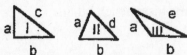

a. Decreasing the right angle in triangle I gives triangle II in which $d < c$. So $a^2 + b^2 > d^2$.

b. Increasing the right angle in triangle I gives triangle III in which $e > c$. So $a^2 + b^2 < e^2$.

75. a. For m, an even number greater than 2, the longest side is $(m^2/4) + 1$ and because
$m^2 + [(m^2/4) - 1]^2 = m^2 + m^4/16 - (2m^2/4) + 1 = m^4/16 + (2m^2/4) + 1 = [(m^2/4) + 1]^2$ the statement is **true**.

b. $(2n + 1)^2 + (2n^2 + 2n)^2 = 4n^4 + 8n^3 + 8n^2 + 4n + 1 = (2n^2 + 2n + 1)^2$, so the statement is **true**.

c. $x^2 + (x + 1)^2 = (x + 2)^2$ is equivalent to $x^2 + (x + 1)^2 - (x + 2)^2 = 0$ which is also equivalent to $x^2 - 2x - 3 = 0$. The solutions to this last equation are 3 and –1. So the only Pythagorean triple with consecutive integers is 3, 4, 5 because –1 cannot be the length of the side of a triangle.

77. The figure is a right triangle because the radius and the tangent meet at a right angle. The shorter leg is the radius of the Earth or 4000 miles. The longer leg is the radius of the Earth plus the height of the mountain. The height of the mountain is 10,560 feet or 2 miles. Thus solving $4002^2 = 4000^2 + x^2$ where x is the distance from the top of the mountain to the point on the horizon, we get $x = 127$ miles.

79. Draw a right triangle between the two hikers. The longer leg is 8 miles + 4 miles = 12 miles long, and the shorter leg is 4 miles long. Let x be the distance between the two hikers and solve $x^2 = 4^2 + 12^2 = 160$; $x \approx 13$ miles.

81. The area of the trapezoid, A, is the sum of the areas of the 3 triangles. So $A = (1/2)ab + (1/2)ab + (1/2)c^2$. The area of a trapezoid is also $(1/2)$altitude(base 1 + base 2). So $A = 1/2(a + b)(a + b) = 1/2(a^2 + 2ab + b^2)$. Thus $(1/2)ab + (1/2)ab + (1/2)c^2 = 1/2(a^2 + 2ab + b^2)$ and $a^2 + b^2 = c^2$.

SECTION 10.5

1. Responses will vary. They should include: pairs of segments AC and BD and AB and CD are parallel and congruent; diagonal segments AD and BC are congruent; E is midpoint of both segment AD and segment BC; pairs of triangles ACD and BCD, CED and BED, AEC and DEB are congruent; pairs of angles BAC and BDC, ABD and ACD, BAD and CDA, ABC and BCD, AEC and BED, AEB and CED are congruent; pairs of angles BAC and ACD, ACD and CDB, CDB and DBA, DBA and BAC are supplementary.

3. In addition to the relations of #1, the diagonals AD and CB are perpendicular and segments AC, CD, DB, and BA are congruent.

5. Rhombus.

7. Square.

9. The quadrilateral is not necessarily any of the listed figures. It could be a general quadrilateral.

11.   The diagonals of all parallelograms bisect each other.

13. The figure is a square.

15. Not possible. A rectangle with all sides congruent is a square. But a square is a rhombus and a rhombus has perpendicular diagonals.

17. Not possible. The diagonals of a rhombus are perpendicular.

19. a. No, it could be a rhombus.   b. No, it not need be a parallelogram

c. No, it not need be a parallelogram   d. Yes

e. No, it could be a parallelogram without right angles. f. No, it could be a rhombus.

21. a, b, c, d, e, f: No.

23. Answers will vary. Possible answers are:
    a. There are two pairs of roads between corners that are the same length.
    b. Roads joining opposite corners are perpendicular to each other.
    c. Roads joining opposite corners could be, but are probably no the same length.

25. The diagonals of a rectangle are congruent.

27. A quadrilateral with opposite sides congruent is a parallelogram. The opposite sides (bar and seat) of a parallelogram are parallel.

29. Responses will vary. Possibilities include: the diagonals of a rhombus are perpendicular; all sides of a rhombus are congruent.

31. Draw a segment *AB*, and construct its perpendicular bisector. Then choose points *D* and *E* on the perpendicular bisector, and draw kite *ADBE*. It is a kite because it has two pairs of adjacent congruent sides, AD ≅ DB and AE ≅ BE, since any point on the perpendicular bisector of a segment is equidistant from the endpoints of the segment.

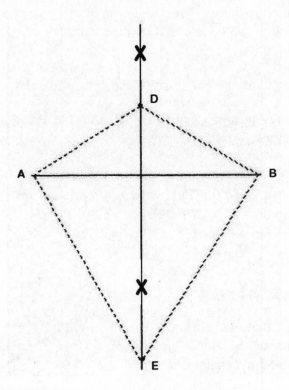

33. Ensures parallelogram.

35. Could be a trapezoid.

37. Ensures parallelogram.

39. Ensures a parallelogram.

41.

43. Since opposite sides of a parallelogram are parallel and the remaining sides serve as transversals, and since same side interior angles are supplementary, the two angles consecutive to the right angle are also right angles. Applying the same argument to the remaining angles, we see that all angles are right angles and the figure is a rectangle.

45. A parallelogram with a right angle has all right angles (see #41). Since the opposite sides of a parallelogram are congruent, the sides opposite to the congruent adjacent sides are congruent to these and to each other. So all four sides are congruent and the figure is a rhombus. A square is a rhombus that is a rectangle.

47. No. Consider a generalization such as, "A rectangle can have a pair of adjacent sides that are not congruent." This generalization would not be true if the word "rectangle" were replaced with the word "square."

49. Responses will vary. A possibility is: The diagonals are perpendicular.

51. Student GES activity. The results are: a. Square. b. Rhombus. c. Parallelogram. d. Rectangle. e. Rectangle.   f. Parellelogram.   g. Parallelogram.

53. GES activity. The length of the segment is half the sum of the lengths of the parallel sides.

55. Student activity.

57. 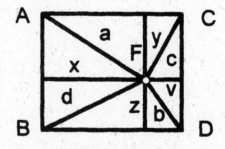 $2x + 2y = 360$. $x + y = 180$. Thus x and y are supplementary.

59. Some students may suggest a proof similar to:
(a) Let F be any point inside the rectangle. Through F draw parallels to the sides of the rectangle. Now:
$a^2 = x^2 + y^2$, $b^2 = v^2 + z^2$, $d^2 = x^2 + z^2$, and $c^2 = v^2 + y^2$.
Further $a^2 + b^2 = x^2 + y^2 + z^2 + v^2$ and
$c^2 + d^2 = x^2 + z^2 + y^2 + v^2$.
(b) So, no matter where F is placed inside the Rectangle, $a^2 + b^2 = c^2 + d^2$.

61. GES activity, a. GES gives a ratio of 5:1.   b. Question will vary. A possibility is: "What if the vertices are connected to trisection points?"

63. Square.

65. Rhombus.

67. Pentagon.

69. A *triquad* is a figure formed from a triangle and a quadrilateral such that the 2 figures share either a common vertex or a common side.

71.

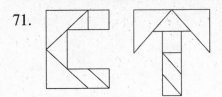

73. Responses will vary.

CHAPTER 10 REVIEW EXERCISES

1. 1, 1, 2, 3, 5. The ratio of a term to the preceding term approaches the Golden Ratio as the number of terms increases. The Golden Ratio is the ratio of length to width in an aesthetically pleasing rectangle.

3.

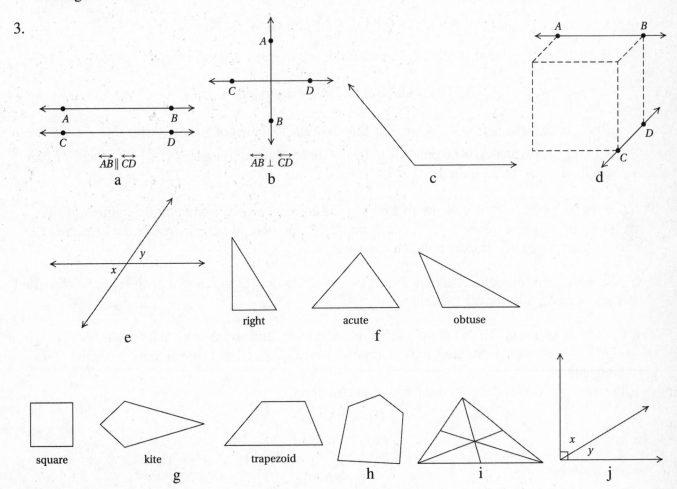

5. Some responses will vary. Possibilities include:
    a. Pairs of non-vertical congruent angles include: 2 and 6 (corresponding), 3 and 5 (alternate interior), 1 and 7 (alternate exterior), 3 and 7 (corresponding).
    b. Pairs of non-adjacent supplementary angles include: 4 and 5, 3 and 6, 1 and 8, 2 and 7.
    c. Pairs of alternate interior angles are: 3 and 5, 4 and 6.

d. Pairs of alternate exterior angles are 2 and 8, 1 and 7.

e. Pairs of corresponding angles are 2 and 6, 3 and 7, 1 and 5, 4 and 8.

7. m (central angle) $= \dfrac{360}{8} = \mathbf{45\,^\circ}$

m (interior angle) $= \dfrac{180(8-2)}{8} = \mathbf{135\,^\circ}$

m (exterior angle) $= 180 -$ measure of interior angle $= \mathbf{45^\circ}$.

9. The sum of the interior angles of any convex pentagon is $(5-2)180^\circ$, so the sum of the interior angles of a house shaped pentagon is **540°**.

11. $m < P = (1/2)(130 - 35)$ and $m < Q = (1/2)(130 + 35)$. So $Q - P = 35/2 - (-35/2) = \mathbf{35^\circ}$.

13. The Euler line is the line that contains the centroid, orthocenter, and circumcenter of the triangle.

15. $h^2 = 10^2 - 6^2$, $h^2 = 64$, $h = 8$. The ladder is 8 ft above the ground.

17. a. $20\sqrt{2}$, or approximately 20.28 cm. The length of the diagonal of a square with side s is $\sqrt{2}\,s$.

 b. $5\sqrt{3}$, or approximately 8.66 cm. The height of an equilateral triangle is $(s\sqrt{3})/2$, where s is the length of a side of the triangle.

19. Since the sum of the exterior angles of a polygon is constant at 360° no matter the number of sides and since there is an exterior angle for each side, as the number of sides increases the measure of each exterior angle of a regular polygon decreases.

21. Yes, it does have four segments each joined to two others only at the endpoints. It could be classified as a non-simple, non-convex quadrilateral.

23. a. Responses will vary. The angles could be cut from paper and arranged so that the sum is a straight angle. A ray could be rotated, moved, rotated, moved, rotated and shown to have rotated 180 degrees.

 b. The same processes could be used with a quadrilateral.

SECTION 11.1

1. One must know the **direction** and the **distance** of the slide.

3. One must know the **point** about which to turn, the direction of the turn, and the turn **angle**.

5. One must know the **line** about which to flip the plane.

7. A reflection. The line of reflection would be horizontal, midway between the tips of the two figures.

9. A reflection followed by a translation (or a translation followed by a reflection). Reflecting in a vertical line between the two figures would change orientation. Then a translation would map the figure onto the final image.

11.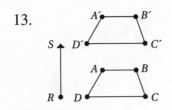

13.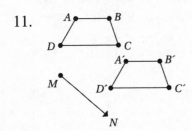

15.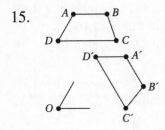

17.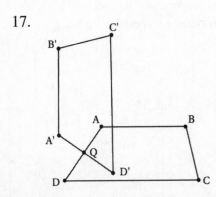

19.

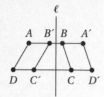

21.

23. a. Because the propeller will superimpose upon itself when rotated 120° or a multiple of 120° about its center, it has rotational symmetry in multiples of 120° about the center.

   b. Because the flag has the elongated cross in the center with the four quadrants the same, it has horizontal and vertical reflectional symmetry. Also, since the arms of the cross are different lengths, it has only 180 degree rotational symmetry about its center.

   c. A half turn will superpose the figure upon itself so it has rotational symmetry in multiples of 180° about its center.

25. A, B, C, D, E, H, I, K, M, O, T, U, V, W, X, Y

27. H, I, N, O, S, X, Z

29. The pattern has a horizontal and a vertical line of reflectional symmetry. It has 180° rotational, or point symmetry.

31. a. The pair is topologically equivalent because the W can be transformed into the spiral without any cutting. Just stretch the W into a segment and then form the spiral.

   b. The two figures are not topologically equivalent. The curved figure has a vertex of degree 4 that cannot be reduced by smoothing. The segmented figure has no vertex of degree 4. If the curved figure is cut at the 'leaves' to separate the 'stem and the roots', then a figure topologically equivalent to the segmented figure is produced.

33. The following sets contain topologically equivalent letters of the alphabet because the elements of a set may be transformed one into another without cutting: $\{A, R\}$, $\{C, I, L, M, N, S, U, V, W, Z\}$, $\{E, F, G, Y, T\}$, $\{H, I\}$, $\{K, X\}$, $\{O, D\}$, $\{P, Q\}$. For example, all the elements of the second set can be 'straightened out' into segments and then reformed into any of the other elements.

35. By definition two figures are similar if there exists a combination of isometries and size transformations such that one figure is the image of the other.

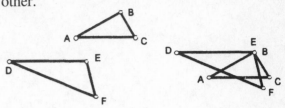

a. In (a), for example, after deciding that, if the figures are indeed similar, *D* would correspond to *A*, *E* to *B*, and *F* to *C*, one could translate one of the figures until one pair of the corresponding vertexes are coincident. Then, if angles *CBF*, *ABD*, and the smaller of the angles formed by the intersection of *DF* and *AC* are congruent, one of the triangles can be rotated to make pairs of sides parallel and the triangles are similar.

b. The same approach may be taken in b after reflecting *EFGH* to establish the same orientation as *ABCD*.

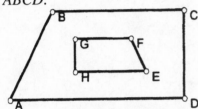
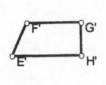
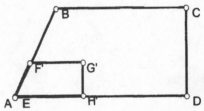

37. Responses will vary. Possibilities include:

TRANSLATION	REFLECTION	ROTATION
a. No change in orientation.	There is a left-right exchange but not a top-bottom.	180 degree rotation will effect both left-right and top-bottom.
b. No points map into themselves.	Points in which the mirror intersects the figure map into themselves.	If the center is on the figure, then that point maps into itself.
c. Only one parameter.	One parameter, the mirror.	Two parameters, center and angle.
d. Measures unchanged.	Measures unchanged.	Measures unchanged.

39.

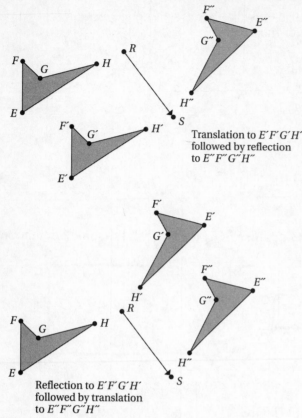

Translation to $E'F'G'H'$ followed by reflection to $E''F''G''H''$

Reflection to $E'F'G'H'$ followed by translation to $E''F''G''H''$

It appears that the order does not matter.

41. a. 180° rotational symmetry.
   b. 120° and 240° rotational symmetry; three lines of reflectional symmetry.
   c. One line of reflectional symmetry.
   d. No symmetry.

43. a. A kite that is not a rhombus has reflection symmetry over its long diagonal.
   b. A parallelogram that is not a rectangle has 180° rotational symmetry about its center.
   c. A rectangle has both rotational symmetry (180° about its center) and reflection symmetry (with respect to lines through the center perpendicular to the sides).

45. a. A scalene triangle.
   b. Answers will vary. Examples are: an isosceles triangle, an isosceles right triangle, an isosceles acute triangle, an isosceles obtuse triangle.
   c. Not possible.
   d. An equilateral triangle.
   e. Not possible.

47. a   A general quadrilateral with no sides congruent or a non-rhombus parallelogram.

b.   Answers will vary. Examples are: an isosceles trapezoid. A kite is not a rhombus.

c.   Answers will vary. Examples are: a rhombus, a rectangle.

d.   Not possible.

e.   A square.

49. a.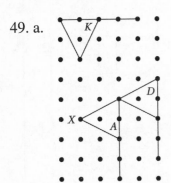

b.   *A* and *K* are congruent because translations and rotations are isometries, that is they preserve measures and relative locations.

51. The rectangle outlines of the two rooms are not similar. For example, if the original room was 10 by 12, the new room would be 14 by 16. In this case the ratios of corresponding sides, 10/14 and 12/16 are not equal and hence the rectangles are not similar.

53. No. The model could be a square, but the garden could be a non-square rhombus (or vice versa). The side lengths could be in the ratio 1:25, but the angles could be different so the shapes would be different.

55. Responses will vary. One possibility is:

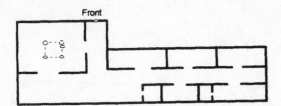

For the house plan shown, sticking to either the left or the right wall will guide a person from the front door to the front door and through every room of the house *providing* there are no 'free-standing' rooms such as the dashed room in the floor plan. Freestanding rooms have no connection to the outer walls of the structure.

Most conventional houses are topologically equivalent to straight lines (i. below). The straight line, in turn, is topologically equivalent to the arc (ii.), to the open rectangle (iii.), and to the 'house' (iv.) in which the outer wall has been stretched and folded to produce the inner walls. So: as one will obviously proceed from *F* to *F* by sticking to either the left-hand or right-hand wall in (ii.), one will also do the same in the house. The inclusion of a free-standing wall, *m* in (v.), does not change the basic topological equivalences.

i.   ii.   iii.   iv.   v.

57. Responses will vary. Since we are concerned only with patterns in which there are 4 boxes in a row, we can eliminate any of the 35 patterns lacking this characteristic. Since the pattern must have rotational symmetry we can further eliminate any pattern with 4 squares in a row and the remaining 2 on the same side of the 4. It is easily seen that patterns with the 2 squares on the same one of the lined squares do not have rotational symmetry. Thus we are left with 2 patterns with the desired characteristics:

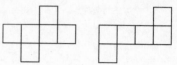

59.

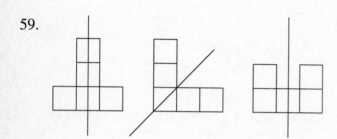

61. The programs work. New figures will vary.

63. The answers are: **a. Yes; b. No; c. No; d. Yes**. Supports for these conjecture will vary. Possibilities include:

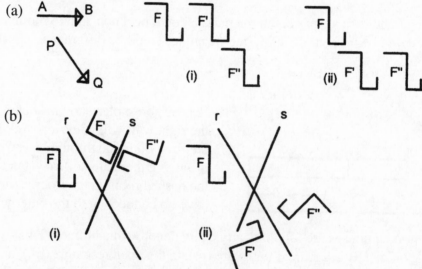

SECTION 11.2

1.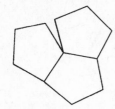

3. **6**, because the angles of an equilateral triangle are 60 degrees and 360/60 = 6.

5. **4**, because the sum of the angles of a quadrilateral is 360°.

7. **Three** regular polygons will tessellate: equilateral triangle, square, regular hexagon.

9.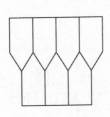

11. **No**, because regular polygons are not used in the tessellation.

13. **No**. Demonstrations will vary.

15. Descriptions and drawing will vary. The equilateral triangles would tessellate the plane, but the colors of the parts of the triangles would make it appear as a tessellation of regular hexagons.

17. Both pentagons will tessellate as shown in the figures.

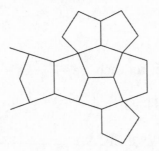

19. Since in *B* each vertex is surrounded by 3 triangles and 2 squares, it may be characterized as 3, 3, 3, 4, 4 or as $3^3 4^2$. In *D* each vertex is surrounded by an equilateral triangle, 2 squares, and a hexagon. So this tessellation may be described as 3, 4, 4, 6 or $34^2 6$. In like manner, *G* is 4, 6, 12.

21. The tessellation is not regular because the tessellating polygon, a hexagon, in itself is not regular.

23. Although the tessellation is accomplished with regular polygons, squares and equilateral triangles, the arrangements at all vertices are not identical, Thus the tessellation is not semi-regular.

25. Because when you try to extend the arrangement to cover the plane, you are forced to have more than one type of arrangement around vertices, as illustrated by points A and B.

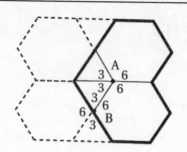

27. The figure has no symmetry.

29.

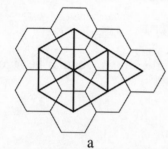

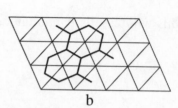

a   b

c. It appears that the tessellations are duals of each other.

31.

    a. Equilateral   b. Square: cut   c. Quadrilateral:
        triangle: cut   and rotate   cut and turn.
        and turn.   at corners.

33. Responses will vary. Possibilities for each are:
    a. TO SQUARE: SIDE   b. TO EQTRI: SIDE
        REPEAT 4[FD: SIDE RIGHT 90]   REPEAT 3[FD: SIDE RIGHT 120]
        TO TESSQ   TO TESSTRI
        PENUP BACK 80 PENDOWN LEFT 90   PENUP BACK 80 PENDOWN RT 30
        REPEAT 4 [SQUARE 30 FD 30]   REPEAT 4[EQTRI 30 FD 30]
        LEFT 180 REPEAT 4[SQUARE 30 FD 30]   LEFT 120 FD 30 RT 60
                                               REPEAT 4 [EQTRI 30 FD 30]

35. All pentominos tessellate. One example is:

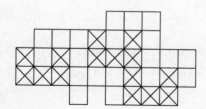

37. A possible heptiamond tessellation is:

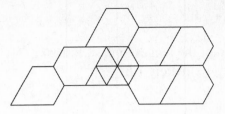

39. 3, 3, 6, 6 and 3, 6, 3, 6

41. 3, 3, 3, 3, 3, 3 and 3, 3, 4, 12 and 3, 3, 4, 3, 4

43. Demiregular, since it has two different vertex arrangements, 3, 12, 12 and 3, 4, 3, 12.

45. Answers may vary. The technique involves making the two new triangles created from the opposite sides of the parallelogram congruent.

47. Two possible tessellations using squares and equilateral triangles are shown below.

49. Description will vary. The tessellation is formed from regular pentagons and regular decagons, 5-pointed stars, and figures made of 2 8-sided portions of decagons fitted together to form a 16-sided concave polygon.

51. Responses will vary.

53. The area of a grid line square is c^2. The area of the larger square in the tessellation is b^2 and the area of the smaller is a^2. Now, 1 + 2 + 3 + 4 make up a square of area b^2, Thus $a^2 + b^2 = c^2$.

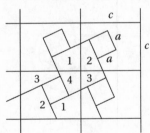

SECTION 11.3

1. The ratio of the length to the width of a 3×5 card is approximately 1.67, the closest to the golden ratio, approximately 1.618.

3. Student activity.

5. Responses will vary. The ratio of the lengths of segments *TH* and *HG* and the ratio of the lengths *TQ* and *TS* is 1.618:1.

7. The star polygon {5/2} has 5 sides: *AC, CE, EB, BD*, and *DA*.

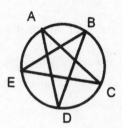

9.

    a. A segment is   b. A quadrilateral is   c. A triangle is   d. A hexagon results.
        formed.   produced.   formed.

11. a. Since *ABCDE* is a regular pentagon, angle *x* measures $(5 - 2)(180)/5 = 108°$ and angle *y*, the supplement of *x*, measures 72°. Since the triangle *BCF* is isosceles, angle *z* measures 72°, and the angle at the point of the star measures $(180 - 72 - 72) = \mathbf{36°}$

    b. The 8 points on the circle divide the circle into 8 equal arcs of 45°. Each point angle is an inscribed angle intercepting two of the 45° arcs. Since an inscribed angle has ½ the degree measure of the arc it intercepts, the point angles have measures of **45°**.

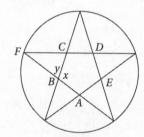

    c. The 9 points on the circle divide the circle into 9 equal arcs of 40°. Each point angle is an inscribed angle intercepting five of the 40° arcs. Since an inscribed angle has ½ the degree measure of the arc it intercepts, the point angles have measures of **100°**.

13. A star polygon, with 8 sides.

15. A star polygon, with 12 sides.

17. Responses will vary. A possibility is

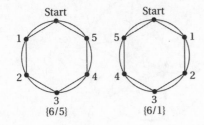

19.

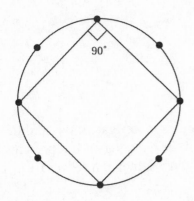

21. Responses will vary. They should include: Star polygons have fewer vertices and sides than do star shaped polygons of the same number of point angles.

23. p(point) = d(dent) – 360/n. $p = 110 – 360/6 =$ **50 degrees**.

25. p(point) = d(dent) – 360/n. $p = 75 – 360/10 =$ **39 degrees**.

27. d(dent) = 360/n + p(point). $d = 20 + 360/5 =$ **92 degrees**.

29. d(dent) = 360/n + p(point). $d = 40 + 360/9 =$ **80 degrees**.

31. *ABCDEFGH* is a regular octagon with interior angles of 135°. All the triangles at the points are isosceles triangles congruent to triangle *GHI* which has a vertex angle of 15°. So the base angles of *GHI* are (180 – 15)/2 = 82.5°. Since all three angles about point *H* sum to 360°, 360° = d + 82.5 + 82.5 + 135 and d, the dent angle, is equal to **60°**.

33. Responses will vary. **No, she was not correct.**
If the point angle is 30°, the dent angle, *p*, is
(360°/6) + 30°, or 90°. Since 90 degrees is the
angle of a square, a square can fill the spaces at
A, *B*, *C*, *D*, *E*, and *F*. However, the angle *JIK* is
140 degrees, far greater than the point angle of
the star polygon. So the pattern cannot be
extended.

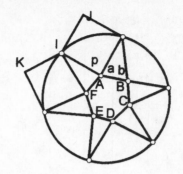

35. To be a star polygon, {*n/d*}, we must have 1 < *d* < *n* − 1, *d*
relatively prime to *n*, and recognize that {*n/d*} is equivalent
to {*n/n* − *d*}. So for *n* = 10, candidates for *d* are 2, 3, 4, 5, 6,
7, and 8. Of these only 3 and 7 are relatively prime to 10.
Since {10/3} is equivalent to {10/7}, there is **one star
polygon with 10 sides**.

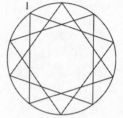

37. Responses will vary depending upon the software used and the figures investigated. Possibilities
include: The generalization given is: If *d* represents the measure of the dent angles, *a* represents the
measure of the point angles, and *n* represents the number of point angles, then *d* = (360/*n*) + *a*. The
figures support the generalization.

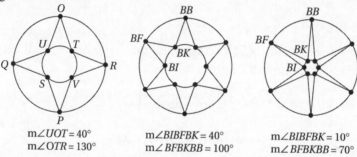

m∠*UOT* = 40°
m∠*OTR* = 130°

m∠*BIBFBK* = 40°
m∠*BFBKBB* = 100°

m∠*BIBFBK* = 10°
m∠*BFBKBB* = 70°

39. Responses to this construction activity will vary. One approach is: Construct a circle with 2
perpendicular diameters. Bisect the right angles at the center of the circle giving eight points equally
spaced on the circle and eight radii joining these points to the center of the circle. Use the eight radii
as the bisectors of 45 degree angles with vertexes the eight points on the circle. The segments joining
the points on the circle to the intersection points of consecutive angle sides are the sides of the 8
pointed star-shaped polygon with 45 degree point angles.

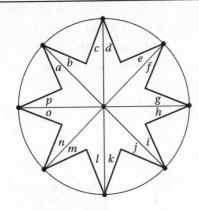

Angles a, b, ... o, p are 22.5 degrees

41. Examples will vary. Possibilities include: True: Suppose $n = 8$ and $d = 6$. Then $n/n-d$ / $d/n-d$ is 4/3. {8/6} represents the figure for {4/3}, a 4-gon, or square. True: Suppose $n = 9$ and $d = 6$. Then $n/n-d$ / $d/n-d$ is 3/2. {9/6} represents the figure for {3/2}, a 3-gon, or equilateral triangle.

43. In Exercise 36 we arrived at the conjecture that the sum of the measures of the point angles of a non-symmetric five-pointed star is 180°. Recall that the sum of a complete set of exterior angles of any concave polygon is 360°. So the sum of the measures of angles a, e, d, c, and b is 360° and the sum of the measures of angles g, h, f, j, and f is 360°. These two sums added to the sum of the measures of the point angles p, q, r, s, and t is 5(180°) = 900° because these 15 angles are all the angles of 5 triangles. Finally, subtracting out the 2 sums, each 360° is of the two sets of exterior angles, we have that the sum of the point angles is 180°.

45. The diagram at the right shows one of the vertices shared by two 9-gons and the 6 pointed star. From this we see that the point angle is **80 degrees**. Now, first construct the 6 pointed star polygon by drawing a circle with diameter d equal to the desired width of the star as measured from star vertex to opposite star vertex. At the center of the circle draw 3 diameters at 120° to each other. Let each of the 6 radii, for example OA, OB, …, be bisectors of the 80 degree angles by constructing 40° angles to each side. The 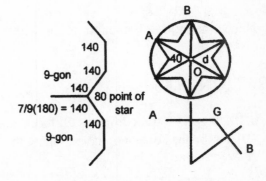 dent angles of these stars measure 80 + 360/6 = **140°**. To construct the 9-gon, construct the perpendicular bisectors to 2 consecutive sides of the star. Use the point of intersection of these bisectors as the center of a circle passing through the 2 points and dent vertex. On this circle mark off 6 more points spaced equally with respect to the point/dent vertex distance. Joining these points in order will produce the 9-gon. Note that the tessellation may be produced with only the 9-gon.

47. Consider the rectangle with width, w, equal to 1 unit and length, L, such that $L/w = (L + w)/L$. So $L/1 = (L + 1)/L$ and $L^2 = L + 1$. Applying the quadratic formula to $L^2 - L - 1 = 0$ we have $L = [-(-1) \pm \sqrt{[(-1)^2 - 4(1)(-1)]}]/2(1) = [1 \pm \sqrt{5}]/2$. Since L is positive, $L = [1 + \sqrt{5}]/2$.

49. Responses will vary. One method is: First construct a circle with 3 diameters at angles of 60°. This circle now has 6 radii. Let each of the radii be the bisector of a 60° angle. Sides of these angles from consecutive points on the circle will intersect at the vertices of the dent angles.

51. Responses will vary. They may include: The tessellation is formed from 10-pointed star polygons, 5-pointed star polygons, and hexagons. The voids within the patterns of these figures are concave hexagons. Now, the point angles of both star polygons are equal ($a = b$) and the hexagons have an angle, d, equal to the dent angle of the 10-point stars with 2 sides of length equal to the sides of the 10 point star. The remaining 4 sides are equal in length to the sides of the 5 pointed star. Two angles of the hexagon, x and y, are equal to the dent angles of the 5 point star. Angles c e, and f are

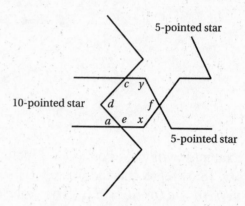

supplements of the point angles of the 10 point star. Now: since the 5 point star is a *pentagon star*, $b = 36°$ and $x = 170 - 72 = 108°$. So the pentagon star has point angles of 36° and dent angles of 108°. The 10 point star has point angles of 36°, dent angles of 72°. The hexagons have one angle of 72°, 3 of 144°, and 2 of 108°.

53. The length of the golden rectangle, DC, is $l + 1/2$, and the width is 1. By the Pythagorean theorem,

$$l = \sqrt{1 + \frac{1}{4}} = \sqrt{\frac{5}{4}} = \frac{\sqrt{5}}{2}$$

Hence the ratio of the length to the width could be expressed in simplified form as $\dfrac{\left(\sqrt{5}+1\right)}{2}$.

SECTION 11.4

1. a.   A regular tetrahedron has all faces equilateral triangles, and all edges congruent. A non-regular tetrahedron does not have to meet these conditions.

    b.

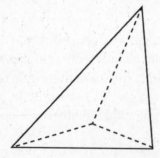

3.   Responses will vary. They might include: a common die, a cereal box, a pencil gripper, a gemstone, a soup can, the tip of a pencil, a softball.

5.   (a) three (b) any quadrilateral prism, right or oblique (leaning)

7. | Polyhedron | F | V | $F+V$ | E |
    |---|---|---|---|---|
    | Triangular prism | 5 | 6 | 11 | 9 |
    | Pentagonal pyramid | 6 | 6 | 12 | 10 |
    | Rectangular pyramid | 5 | 5 | 10 | 8 |

Since $F + V = E + 2$, Euler's formula holds.

8. 10 vertices, 7 faces, 15 edges. $10+7=15+2$, so the formula works.

9. 7 vertices, 7 faces, 12 edges. $7+7=12+2$, so the formula works.

11. 24 vertices, 14 faces, 36 edges. $24+14=36+2$, so the formula works.

13.   (a) The figure has one horizontal plane of reflectional symmetry and four vertical planes of reflectional symmetry, each through a vertex of the equilateral triangular face and the midpoint of the side opposite that vertex. (b) The figure has five planes of reflectional symmetry, each containing the vertex of the pyramid and the midpoint of a side of the pentagonal base. (c) The figure has four planes of reflectional symmetry, each containing the vertex of the pyramid and the midpoint of a side of the square base.

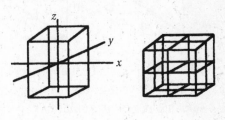

15. Three axes of rotational symmetry, one through each pair of opposite faces. Three planes of reflectional symmetry, one horizontal and two vertical.

17. A right pentagonal pyramid has 5 planes of symmetry.

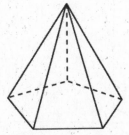

Right Pentagonal Pyramid,
with 5 planes of symmetry

19. It must be a right regular prism, with a base that has rotational symmetry.

21. Trial shows that patterns *a and b* can be folded to form open-top boxes.

23. a.   b.

25. a. Slicing the cone produces an ellipse.
    b. Slicing the cylinder produces a rectangle.

27.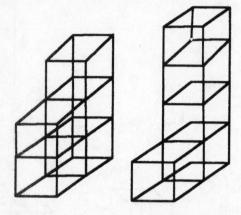

29. This problem is best approached with a light source that produces parallel rays of light, for example a slide projector, and a cube. Shadows that may be seen include a square, rectangles, a regular hexagon, and irregular hexagons.

31. Responses will vary. A possibility is: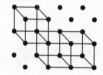

33.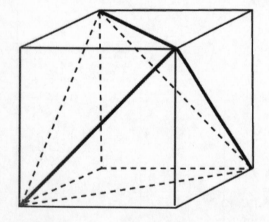

    The edges are all diagonals of the faces of the cube, and so are congruent.

35. Answers will vary. A possibility is: The symmetry of the regular polyhedra give each face the same probability of falling up or down.

37. a. For the square antiprism $F = 10$, $V = 8$, $E = 16$, and $F + V = 18$. Since $F + V = E + 2$, Euler's theorem holds.

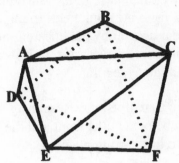

   b. For the triangular antiprism, we may first envision 2 triangles, *ABC* and *DEF* in parallel planes. Each vertex of the top triangle, *ABC*, is joined to 2 points of the lower triangle producing 3 faces, and each vertex of the lower triangle is joined to 2 points of the upper triangle producing 3 more faces. Combined with the upper and lower triangles as faces, we have $F = 8$, $V = 6$, and $E = 12$. Since $F + V = E + 2$, Euler's theorem holds.

39.

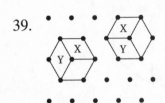

41. Answers will vary. A possibility is: When a corner is sliced off a prism there is a gain of $n - 1$ vertices, a gain of 1 face, and a gain of n edges. So there is a change of $(n - 1) + 1 = n$ in $V + F$ which is equal to the change in E. Euler's formula holds.

43. The figure formed, as shown, is a hexagon. If a coordinate system is superimposed on the cube with side s, the vertices of the hexagon *KLMNPQ* are $(0, s, s/2)$, $(s/2, s, s)$, $(0, s/2, 0)$, $(s, s/2, s)$, $(s/2, 0, 0)$, and $(s, 0, s/2)$ respectively. Each of the sides of the hexagon is the hypotenuse of an isosceles triangle with legs of length $s/2$ and has a length of $(s/2)\sqrt{2}$. Now, if the triangles *MKL*, *KLN*, *LNQ*, *NQP*, *QPM*, and *PMK* are congruent, then the interior angles are congruent and the hexagon is regular. Applying the distance formula to the third side of each of the 6 triangles [for example: triangle *MKL*. $ML =$

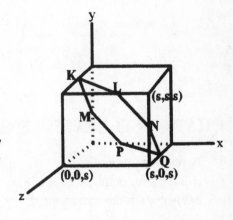

$\sqrt{[(0 - s/2)^2 + (s/2 - s)^2 + (0 - s)^2]} = s\sqrt{(3/2)}$ we find that the lengths are all $s\sqrt{(3/2)}$. So the triangles are congruent and the figure is a **regular hexagon**.

45. a. In order to fit about an edge with no gaps or overlaps, the sum of the dihedral angles of the polyhedra must be 360°. The cube, with a dihedral angle of 90°, is the only polyhedron with a dihedral angle that is a factor of 360 and thus the only polyhedron that will fill the space about an edge.

   b. A combination of 2 tetrahedra and 2 octahedra, angle sum 2(70°32') + 2 (109°28') = 360°, will fill the space about an edge.

47. a.   The regular polyhedra that are deltahedra are **tetrahedra, octahedra, and icosohedra**.
    b.   The double triangular pyramid is a deltahedron because all faces are equilateral triangles but it is not regular as seen by the different number of edges at different vertices. A tetrahedron is a deltahedron that **is** regular.

49. No, the drawings are illusions. Responses will vary. They may include statements as: A vertical support cannot be both at the front and the back of a structure; or, by continually going down steps I cannot return to my starting position.

51. Responses will vary. A possibility: A cube has 8 vertices, 12 edges, and 6 faces. So, for a cube, edges + 2 = faces + vertices. Now, cutting off a corner replaces 1 vertex with 3, and adds a face. So the sum of the faces and vertices increases by 3. But the cut produces three additional edges for the added triangular face. So the difference between $F + V$ and E is still 2 and Euler's Theorem still holds.

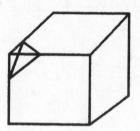

53. b.   An octahedron is formed
    c.   A tetrahedron is formed

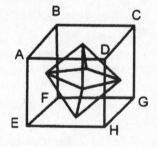
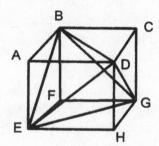

CHAPTER 11 REVIEW EXERCISES

1.   A regular tetrahedron has 4 congruent equilateral triangles as faces; a cube has 6 congruent squares as faces; an octahedron has 8 equilateral triangles as faces; a dodecahedron has 12 regular hexagons as faces; and an icosahedron has 20 equilateral triangles as faces.

3.   A pyramid has 5 faces, 5 vertices, and 8 edges. $8 + 2 = 10$ ($V + F = 10$, $E = 8$).

5.

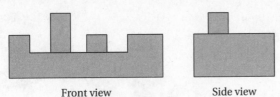

Front view   Side view

Top view

7. Drawings will vary.

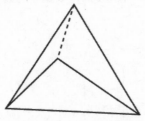

9. Dent angle = 360/ # of points + point angle = 360°/8 + 36° = 45° + 36° = 81°.

11. 3 planes of symmetry (two vertical planes through the midpoints of the parallel bases, and one horizontal plane, through the midpoints of the vertical sides), 3 axes of rotational symmetry, each through a centerpoint of a face of the prism.

13. a. The tessellation is made from two types of regular polygons with the same arrangement at each vertex.
    b. The arrangement of the polygons is not the same at each vertex.

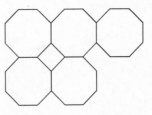

15. a. There are an infinite number of possibilities. Two of these are:

    b. Two possibilities are:

17. **It is true**. Student support will vary. One possibility is to present several special cases.

19. Because the tessellation is to use *equilateral* triangles, the
    dent angles of the 12-pointed star polygons are 60°. Then
    the points angles are 60 – (360/12) = **30°**.

21. Responses will vary.

23. Of the 12 pentominos, only 8 form open top boxes. But of the 12 pentominos only 1 has no more
    than 2 squares in a row and this does have reflection symmetry. It also tessellates the plane.

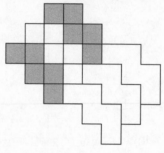

25. Student activity. Responses will vary. The ratio of length to width should be the Golden Ratio.
    Some students may use the difference between the ratios, others the ratio of the ratios. The
    advantages/disadvantages of each may be discussed.

SECTION 12.1

1. Responses will vary. Representative measurements are: The width of a paperback book is about 3 1/4 paperclips and 1 1/5 index fingers.

3. A piano is heavy so kilograms would be most appropriate.

5. Either is appropriate. The measure would be several tens of grams or tenths of a kilogram.

7. The **kilogram** is more appropriate.

9. **kilograms**

11. **milligrams**

13. **grams**

15. 20 pts. = 10 quarts = **2.5 gal**.

17. $12 \text{ ft}^2 = 12 \text{ ft}^2(1/9 \text{ yd}^2/\text{ft}^2) = 12/9 \text{ yd}^2 = \mathbf{1\ 1/3\ yd^2}$.

19. 25 l = 25 l(1000 ml/l) = **25,000 ml**.

21. 2500 cm = 25 m = **0.025 km**.

23. 0.012l = 0.012l(1000 ml/l) = **12ml**.

25. 2000 mg = 2000 mg(0.001 g/mg)(0.001 kg/g) = **0.002 kg**.

27. $20 \text{ cm}^2 = 20 \cdot 100 \text{ mm}^2 = \mathbf{2000\ mm^2}$.

29. 3 m: the height of a ceiling.

31. 1 lb: a couple of paperback books.

33. 6 yd: the length of a car.

35. 500 ml: a really good-sized glass of beer.

37. $-15° \text{ C} = \mathbf{5°\ F}$

39. $0° \text{ C} = \mathbf{32°\ F}$

41. $20° \text{ C} = \mathbf{68°\ F}$

43. $-13° \text{ F} = \mathbf{-25°\ C}$

45. 40° F = **4.4° C**

47. 86° F = **30.0° C**

49. A four-door sedan is between 4 and 5 meters long.

51. A gallon of water weighs approximately 3.5 kg.

53. We know that a kilogram is about 2.2 pounds, so 220 lbs/100 kg.

55. There are about 2.54 centimeters in an inch. So a woman who is 5 feet or 60 inches tall, would be about 150 cm tall (actually 152.4 cm).

57. One hundred meters is longer than 100 yards, so it would take longer to run 100 meters.

59–63. Responses will vary. Although the reasonableness of the statements is determined by obtaining Fahrenheit equivalents, students should become sufficiently familiar with the Celsius scale of temperature to make judgements without conversion. Possibilities are:

59. 300 pounds. Explanations will vary.

61. 36 mpg. Explanations will vary.

63. Eighty degrees C is about 176°F. Water boils at 212°F, so 176°F is very hot. The statement is **not reasonable**.

65. 40° C is 40 degrees above freezing on the Celsius scale. Since 1 C° is equivalent to 9/5 F°, 40° C is 72 degrees above freezing on the Fahrenheit scale, or about 104° F. Since this temperature is almost too hot to play tennis, the statement is **not reasonable**.

67. Since, by definition, water freezes at 0° C and –5° C is a lower temperature, the statement is **reasonable**.

69. a. Since the unit of area has an area of 1 in.^{2}, the 8 in. × 10 in. rectangle has an area of **80 units or 80 in^2**.
    b. The unit has an area of 2 in.^{2} So 80 in.^{2} (1/2 unit/in.^{2}) = **40 units**.
    c. The unit has an area of 1/2 in.^{2} So the area is 80 in.^{2} × (2 units/in.^{2}) = **160 units**.

71. By tracing the right trapezoid, cutting it into a triangle and a square, and covering the large trapezoid, the area is **8 units**.

69–73. Responses will vary. Although judgements are validated by conversion to the more familiar English units, students should become sufficiently familiar with the metric units to make judgements directly. Possibilities include:

73. Since 1 m is approximately equivalent to 1.1 yards or 3 1/4 ft., 2 m would be a little more than 6 feet tall. It would be a short player, but the answer is **reasonable**.

75. 50 sq. m would be about 50 sq. gal. or 450 sq. ft.

77. This is **reasonable**. Fifteen meters is about 45 to 50 feet. Old three-story buildings with high ceilings and pitched roofs easily could be 45 m.

79. **–10° F** is colder than –10° C. –10° F $\approx$ –23° C and –10° C < –23° C. The Fahrenheit temperature is 42 degrees below freezing and the Celsius temperature is 10 degrees below freezing. The Celsius degree is almost twice as big as the Fahrenheit but –10° F is 4 times the number of Fahrenheit degree below freezing than is –10° C.

81. Responses will vary. A possibility is: As the unit of measure increases in size the measure of an object decreases in proportion. So if the unit of measure doubles, the measure halves. If the unit of measure would double again, the measure would be 1/4 of the original measure.

83. Since 180 Fahrenheit degrees (the number of degrees between boiling and freezing water) correspond to 100 Celsius degrees (again, the number of degrees between boiling and freezing water), 20 Celsius degrees correspond to **36 Fahrenheit degrees**.

85. $6.20/kg = $6.20/1000 g = $0.0062/g. Now: ($10)/($0.0062/g) = **1612.9 g** to the nearest tenth of a gram.

87. (10 g /sheet)(500 sheet/ream) = 5000 g/ream = 5000 g/ream (1 kg/1000 g) = **5 kg/ream**.

89. Responses will vary. A possibility is: 40 km is about 24 miles or 1 km is about 0.6 miles. So to obtain a distance in miles, knowing the measure in km, multiply the measure by 0.6. Conversely, knowing a measure in miles to obtain a rough distance in km, multiply by 2. If a better estimate is required, after multiplying the number of miles by 2, subtract one third the number of miles. For example: 200 miles is about 400 km. The better estimate is 400 km – 1/3(200)km or about 330 km.

91. Responses will vary. The two primary arguments are: most of the rest of the world uses metric measures and because of the global interactions in manufacturing it is necessary to have a global standard. On the other hand, a massive reeducation effort is required to change. It has been done in other countries. South Africa and Canada are an examples.

SECTION 12.2

1. To find the area of the rectangle 12 cm × 40 mm one might first change the measures to the same units, either 12 cm × 4 cm or 120 mm × 40 mm.

3. Baseboards generally go around a room, so the measure is a **perimeter**.

5. The coverage of paint is given in square ft, so the measurement involves **area**.

7. Perimeter = 4(8 in) = **32 in**. Area = (8 in)(8 in) = **64 in^2**.

9.   Perimeter = 2(300 cm) + 2(22 cm) = **644 cm**. Area = (300 cm) (22 cm) = **6600 sq. cm**.

11.   Perimeter = 2(6 cm) + 2(4 cm) = **20 cm**. Area = (6 cm)(3 cm) = **18 cm²**.

13.   Perimeter = 2(7 in) + 2(3 in) = **20 in**. Area = (3 in)(5 in) = **15 in²**.

15.   Area = (6 in)(6 in) + (1/2)(6 in)(6 in) = **54 in²**.

17.   Area (1/2)(4 m)(10 m+ 13 m) = **46 m²**.

19.   Area = 8(1/2)(5 cm)(6 cm) = **120 cm²**. Perimeter = 8(5 cm) = **40 cm**.

21.   Circumference ≈ 10 mm(3.14) = **31.4 mm**. Area ≈ 3.14[(1/2)(10 mm)]² = **78.5 mm²**.

23.   Circumference ≈ 2(4 in.)(3.14) = **25.12 cm**. Area ≈ 3.14(4 in.)² = **50.24 cm²**.

25.   Areas of similar figures are proportional to the squares of corresponding lengths. And since price is proportional to area we have: $A_1/A_2 = P_1/P_2 = L_1{}^2/L_2{}^2$ where A represents area, P represents price, and L represents length. So $P/\$110 = 6^2/2^2$. $P = (36/4)110 = $ **$990**.

27.   Since the areas of similar polygons are proportional to the squares of perimeters, the perimeters are proportional to the square roots of the areas. Thus $P1/P2 = \sqrt{36}/\sqrt{81} = 6/9 = $ **2/3**.

29.   The figure can be divided into a rectangle 9 cm by 6 cm and a trapezoid with bases 9 cm and 4 cm and height 5 cm. So the area is (9 cm)(6 cm) + (1/2)(5 cm)(9 cm + 4 cm) = 54 cm² + 32.5 cm² = **86.5 cm²**.

31.   Since the legs of a 45/45/90 triangle are equal, the area is (1/2)(12 yd)(12 yd) = **72 sq. yds**.

33.   The base of the triangle is 7.5 m and the height 4 m. So the area is (1/2)(4 m)(7.5 m) = **15 m²**.

35.   a.   The length of the arc is (80/360)(2)(3.14)(5 m) ≈ **6.98 m**.
      b.   The length of the arc is (120/360)(2)(3.14)(10 cm) ≈ **20.9 cm**.

37.   Since the altitude of ADE is ½ the altitude of ABC and the base of ADE is also ½ of the base of ABC, the area of ADE is (1/2)(1/2) the area of ABC. Thus the ratio area ADE to area ABC is **1 to 4**.

39.   The formula for the area of a rectangle is bh, where b is the base and h is the height. As you move to a trapezoid, there are 2 bases, so you are essentially finding the mean of the lengths of the two bases when you use the formula $(1/2)h(b_1+ b_2)$. As you move to a triangle, one of the bases disappears, so you are left with $(1/2)bh$.

41.   Since the ratio of areas is the square of the ratio of perimeters of similar figures and since the ratio of corresponding sides of similar figures is the same as the ratio of perimeters, if the ratio of areas is 3/1, the **ratio of corresponding sides is $\sqrt{3}/1$ or about 1.732:1. The ratio of the perimeter is also $\sqrt{3}:1$**.

43. Since the area of a rectangle is length times width and the area of a square is side squared, we have s^2 = (12 in)(27 in) = 324 sq in. So s = **18 in**.

45. No. Reasons will vary. Consider the following rectangles: 1×10 and 3×5.

47. No. Reasons will vary. Consider the following dimensions: 1×10 and 4×4.

49. False. Reasons will vary. Consider the following dimensions: 1×10 and 4×5.

51. Responses will vary. One possible argument is: The sides given may not be corresponding. One may be the longer side of one of the rectangles and the other the shorter side of the other rectangle.

53. Arguments will vary. A possibility is: **The rectangle becomes more elongated, or less square**. A square has the smallest perimeter for a given area. Suppose the area is 64. Then the perimeter of the square with this area is 32. Now, in order to have an area of 64, the product of the length and with must be 64. Some pairs with this product are 4 and 16, 2 and 32, 1 and 64. The respective perimeters are 40, 68, and 130. As the perimeter increases with constant area the rectangles becomes less square.

55. The area of the dartboard is (6 ft)(7 ft) = 42 sq ft. The area occupied by the balloons, the success area, is $63(3.14)[(1/2)(6 \text{ in})(1 \text{ ft}/12 \text{ in})]^2$ = 12.37 sq ft. Thus the probability of success is 12.37/42 = **0.29**.

57. Both will run the same straight distance, 200 yds. Alan runs a circle, the sum of the 2 semi-circles, with a diameter of d. Billy will run a circle of diameter $d + 6$ because there is an additional 3 yds at each end of the diameter. So Alan will run $200 + 3.14d$ and Billy will run a distance $200 + 3.14(d + 6)$. The difference, the extra distance Billy runs, is $6(3.14)$ = **18.84 yd**.

59. Arguments will vary. One possibility is: Represent the length of DC with x and the length of ED with y. Then the area of the trapezoid $ABCD$ is $(1/2)y(x + 4x) = (5/2)xy$. The area of triangle CDE is $(1/2)xy$. Thus the ratio of the areas is **5/1**.

61. Since the diagonal of the larger square is the hypotenuse of an isosceles right triangle with the side, s, of the square as legs, $2s^2 = 144 \text{ mm}^2$. So the area of the larger square, s^2, is 72 mm^2. The area of the smaller square is (6 mm)(6 mm) = 36 mm^2. So the shaded area is (72 – 36) mm^2 = **36 mm^2**.

63. The area of the square is (6 cm)(6 cm) = 36 cm^2. The diameter of the circle is 6 cm, thus the radius is 3 cm and the area is $(3.14)(3 \text{ cm})^2$ = 28.26 cm^2. The shaded area is the difference, **7.74 cm^2**.

65. Responses will vary among groups of students. One dissection is:

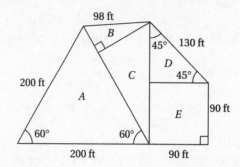

This dissection gives an area of $A + E + D + (B + C) = [(200)^2/4] \sqrt{3} + (90)^2 + (1/2)(90)(90) + (1/2)(90)(200) \approx \textbf{38,470 ft}^2$.

67. Arguments will vary. A possible argument is: **Tony is correct**. Perimeter is not directly related to area. For example, walking along a path along the shore that is concave encompasses less surface water than walking the same distance along the share on a path that is convex.

69. The owner envisioned a pool that was similar in shape to the pool of the competitor but with each linear dimension twice as large; that is a pool 10 m by 14 m with an area of 140 sq m. The contractor considered the area to be the important characteristic of the pool and constructed a pool with twice the area. If the shape of the new pool is similar to the competitor's pool, its dimensions are $5\sqrt{2} \times 7\sqrt{2}$ or about 7.07 m × 9.9 m. The owner would consider this pool less than 50% larger (actually, about 41%).

71.

No. of Sides	Length One Side	Perimeter	$P/D, D = 2$
48	0.130806256	6.278700299	3.139350149
96	0.065438165	6.282063794	3.141031897
192	0.032723463	6.282904837	3.141452419
384	0.016362279	6.283115108	3.141557554

SECTION 12.3

1. $V = (5)(5)(5) = \textbf{125 cm}^3$ $SA = 6[(5)(5)] = \textbf{150 cm}^2.$

3. $V = [(1/2)(3)(3)]6 = \textbf{27 mm}^3$ $SA = 2[(1/2)(3)(3)] + 2[(3)(6)] + 3\sqrt{2}\,(6) \approx \textbf{70.5 mm}^2$

5. $V = (3)(5)(12) = \textbf{180 in}^3$ $SA = 2[(3)(5)] + 2[(3)(12)] + 2[(5)(12)] = \textbf{222 in}^3$

7. $V = 3.14(4^2)4 \approx \textbf{200.96 cm}^3$ $SA = 2[3.14(4^2)] + 3.14(8)4 \approx \textbf{200.96 cm}^3$

9. $V = (1/3)(12)(12)(8) = \textbf{384 cm}^2$ $SA = (12)(12) + 4(1/2)(12)\sqrt{(8^2 + 6^2)} = \textbf{384 cm}^2$

11. $V = (4/3)(3.14)(10^3) = \textbf{4186.67 cu. in.}$ $SA = 4(3.14)(10^2) = \textbf{1256 sq. in.}$

13. The surface area of the figure is the surface area of the cylinder with only one base plus half the surface area of the sphere. The surface area of the cylinder with one base is $32\pi + 16\pi$, and the surface area of the half-sphere is $(1/2)(32)\pi$. So the total surface area is $(32 + 32 + 16)\pi$ or 251.2 cm^2. The volume of the figure is the volume of the cylinder plus half of the volume of a sphere. The volume of the cylinder is $4(4^2)\pi$, and the volume of the half-sphere is $(1/2)(4/3)(4^3)\pi$, or 334.9 cm^3.

15. $V = (1/3)(3.14)(6/2)^2 \sqrt{(6^2 - 3^2)} \approx \mathbf{48.9 \text{ cm}^3}$.

17. $V = (1/3)(3.14)4^2(10) + (1/2)(4/3)(3.14)4^3 = \mathbf{301.4 \text{ cm}^3}$

19. The volume of the rectangular prism is $10 \text{ ft} \times 10 \text{ ft} \times 50 \text{ ft}$ or 5000 ft^3. The volume of the square pyramid is $(1/3)(10 \times 10)(6)$ or 200 ft^3. So the volume of the entire figure is 5200 ft^3.

21. a. $A = 2\pi R^2 + 2\pi RH$; $a = 2\pi (R/2)^2 + 2\pi (R/2)(H/2) = (1/4)A$. So $A/a = \mathbf{4/1}$.
   b. $A = 2(LW + LD + WD)$; $a = 2[(3/4)L(3/4)W + (3/4)L(3/4)D + (3/4)W(3/4)D] = (9/16)A$. So $A/a = \mathbf{16/9}$.

23. $W = (3/4)(4)(3.14)(3850)^2 = \mathbf{139{,}627{,}950 \text{ mi}^2}$.

25. a.   Corresponding linear parts of similar cylinders are the radii, diameters, heights, and circumferences.
   b.   Corresponding linear parts of similar spheres are radii, diameters, and great circles.
   c.   Corresponding linear parts of similar cones are base radii, base diameters, heights, slant heights, base circumferences.
   d.   Corresponding linear parts of similar pyramids are corresponding segments of the bases, heights, and corresponding parts of the faces.

27. Fred is making his tent in the shape of a pentagonal based pyramid. The formula for the surface area of a pyramid is $B + (1/2)ph$ in which B represents the area of the base, p represents the perimeter of the base, and h represents the slant height or the altitude of the triangular faces. So Fred would have to measure the length, l, of a side of the regular pentagonal base to determine p; the distance from the center of the base to the middle of a side, a, to calculate the area of the base; and the distance from the top of the tent to the middle of a side of the base, h. If the tent were to have a dirt floor Fred would not need the measure B and thus would only have to measure l and h.

29. The volume of a sphere is $V = \dfrac{4}{3}\pi r^3$ and a cylinder is $V = \pi r^2 h$. Setting these two equal, we get $\dfrac{4}{3}\pi r^3 = \pi r^2 h$ or $\dfrac{4}{3}r = h$. Hence $h = \dfrac{4}{3}(6) = 8 \text{ cm}$.

31. If the container is cubical, its volume, V, is s^3, s the length of an edge. So $216 \text{ in.}^3 = s^3$ or $s = \sqrt[3]{216} = \mathbf{6 \text{ in}}$. Alternative sized containers will vary. Possibilities are: If the base were 4 in. by 5 in. the height would be $216 \text{ in}^3/20 \text{ in}^2 = 10.8 \text{ in}$. So a container could be 4 in. by 5 in. by 10.8 in. Another possibility is 5 by 6 by 7.2. A third possibility is 3 in. by 5 in. by 14.4 in.

33. The rocks produce a volume increase of $(2 \text{ cm})(30 \text{ cm} \times 60 \text{ cm}) = \textbf{3600 cm}^3$.

35. Arguments will vary. **Tracy is correct.** The original volume, v, is (2 in)(3 in)(5 in) and suppose the order is lwh. Now, the new volume, V, is [(2)(2 in)] [(2)(3 in)][(3)(5 in)]. But because multiplication is both commutative and associative, the multipliers can be associated with any of the dimensions and, in fact, can be factored and themselves multiplied to give $V = 12[(2 \text{ in.})(3 \text{ in.})(5 \text{ in.})]$.

37. The volume, C, of the pipe itself is $(500 \text{ cm})(3.14)(1.75 \text{ cm}/2)^2$, 1202.03 cm^3, and the volume, H, of the hole in the pipe is $(500 \text{ cm})(3.14)(1.5 \text{ cm}/2)^2$, 833.125 cm^3. The volume of the copper is the difference between these volumes, **318.9 cm**^{3}.

39. Let r represent the radius of the tennis balls. The can, a cylinder, also has base radius r and height 6 r. So the volume of the can, C, is $(\pi r^2)(6r) = 6\pi r^3$ and the volume of the 3 balls, B, is $(3)(4/3)\pi r^3 = 4\pi r^3$. So the percent of the can filled by the balls is $(B/c)(100\%) = \textbf{66 2/3\%}$.

41. The concrete fills a bottom slab, $(9 \times 5 \times 2/12) \text{ ft}^3 = 7.5 \text{ ft}^3$; 2 end slabs, each $(4 - 2/12)(5)(2/12) \text{ ft}^3 = 3.19 \text{ ft}^3$, a total 6.39 ft^3; and 2 side slabs, each $(9 - 4/12)(4 - 2/12)(2/12) \text{ ft}^3 = 5.537 \text{ ft}^3$, total 11.07 ft^3. The total of all 5 slabs is **24.96 ft**^{3}.

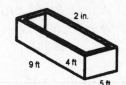

43. A side view of the cylinder and cone is at the right. Triangles BGE and CDE are similar so $50/6 = CD/3$ and $CD = 25$ cm. The volume of the cylinder, C, is $(3.14)(3^2)(25) = 706.5 \text{ cm}^3$. The volume of the cone, V, is $(1/3)(3.14)(6^2)(50) = 1884 \text{ cm}^3$. So, $(C/V)(100\%) = \textbf{37.5 \%}$.

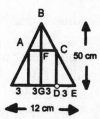

45. The volume of the box, B, is $(6 \text{ in})(9 \text{ in})(6 \text{ in}) = 324$ cu in. The total volume of the 6 cans, T, is $6(3.14)(3/2)^2 6$ cu in = 254.34 cu in. The percent filled by the cans is $(254.34/324)(100\%) = \textbf{78.5 \%}$. Now, if the volume of the box remains at 324 cu in but the dimensions are altered so that it will hold 8 cans of radius r and height h, then the box has dimensions $4r$ by $8r$ by h and $32r^2h = 324$ cu in. The total volume of the 8 cans is $8(3.14)r^2h = 25.12 \ r^2h$. Now, $r^2h = 324/32 = 10.125$ and the volume of the cans is 254.34 cu in. Thus the percent of the box filled by the cans is $(254.34/324) = \textbf{78.5 \%}$.

47. The surface area is $(8 \times 8) + 4(8 \times 1) + 2(3 \times 4) + 2(3 \times 8) + 2(2 \times 8) + (4 \times 8) - 2(3.14)(1^2) + (3.14)(2)(4) = \textbf{251 cm}^2$. The volume is $(8 \times 8 \times 4) - 2(8 \times 2 \times 3) - (3.14)(1^2)(4) = \textbf{147 cm}^3$.

49. The surface area of the first figure is 62.8 cm^2, and the volume is 37.68 cm^3. The surface area of the second figure is 175.84 cm^2, and the volume is 150.72 cm^3. The surface area of the third figure is 100.48 cm^2, and the volume is 75.36 cm^3.
    a. Doubling the radius with the same height more than doubles the area.
    b. Doubling the radius with the same height quadruples the volume.
    c. Doubling the height with the same radius less than doubles the surface areas.
    d. Doubling the height with the same radius doubles the volume.

51. Since four squares of area x^2 are cut from a sheet with an area $(15 \text{ cm})(15 \text{ cm}) = 225 \text{ cm}^2$, the area remaining to make the box is $225 - 4x^2$. The maximum volume (243 in.^3) occurs when $x = 3$.

53. A volume twice that of a unit cube is 2 cu in. So $2 \text{ in.}^3 = e^3$, e the length of an edge. So $2 = \sqrt[3]{2} = \mathbf{1.26 \text{ in}}$.

CHAPTER 12 REVIEW EXERCISES

1. Since $1 \text{ m} = 100 \text{ cm}$, $(1 \text{ m})^2 = (100 \text{ cm})^2 = 10,000 \text{ cm}^2$, **d**.

3. Since there are 9 sq. ft. in 1 sq. yd., there are 200 sq. yd. In the house, **b**.

5. The area of the triangle may be calculated as $(1/2)(20 \text{ in})(15 \text{ in})$ or as $(1/2)(25 \text{ in})(BD)$. So we have: $(1/2)(20 \text{ in.})(15 \text{ in.}) = (1/2)(25 \text{ in.})(BD)$; $BD = \mathbf{12 \text{ in.}, c}$.

7. Suppose the length of CD is x and the altitude of the trapezoid is a. Then the area of the trapezoid is $(1/2)(x + 2x)a = (3/2)xa$. The area of triangle CDE is $(1/2)xa$. So the ratio of the trapezoid area to the triangle area is $[(3/2)xa]/[(1/2)xa] = (3/2)/(1/2)$ is $\mathbf{3/1, a}$.

9. The ratio of the areas of similar figures is the square of the ratio of corresponding lengths. So the area ratio is $(4/1)(4/1) = \mathbf{16/1, c}$.

11. The volume of the cylinder is $\pi(r^2)(4)$. The volume of the cone is $\frac{1}{3}\pi(r^2)(h)$. Equating the two volumes, we have $\frac{1}{3}h = 4$ ah $= 12 \text{ ft}$.

13. a. $SA = 2(6^2/4)\sqrt{3} \text{ sq cm} + 3(6)(10) \text{ sq cm} \approx \mathbf{211 \text{ cm}^2}$.
    b. $V = [(6^2/4)\sqrt{3}]10 \approx \mathbf{156 \text{ cm}^3}$.

15. The original area is 800 ft^2 and the increased area is 1056 ft^2, an increase of 256 ft^2. Thus the percent increase is $(256/800)(100\%) = \mathbf{32 \%}$.

17. Because ECD is similar to ACB and the ratio of corresponding lengths is $1/2$, the area ratio is 1 to 4. So the area of the trapezoid is $3/4$ the area of ACB or 3 times the area of ECB. Thus the ratio of the areas ECD to $AEDB$ is $\mathbf{3/1}$.

19. The area can be dissected into an isosceles right triangle with legs 70 yds and area 2450 sq yd, a rectangle 150 yd by 30 yd with area 4500 sq yd, and a rectangle 80 yd by 70 yd with area 5600 sq yd. So the total area is 12,550 sq yd or 112,950 sq ft or 2.6 acre. The area is closer to 3 acres than to 2 acres but is about 13% shy of the 3 acre claim.

21. The volume of the pool is $(25 \text{ m})(15 \text{ m})(1.5 \text{ m}) = 562.5 \text{ cu m}$. Since there are 1000 l/m^3, the volume is $562,500 \text{ l}$. So the time to fill the pool is $562,500 \text{ l}/ 20 \text{ l/min} = 28,125 \text{ min or } \mathbf{469 \text{ hr or } 19.5 \text{ days}}$.

23. Set the dimensions of a 20″ set be represented by $3K$ and $4K$. Then $(3K)^2 + (4K)^2 = 20^2$. $K = 4$ and the dimensions of a 20 in. set are 12×16 for an area of 192 sq. in. Similarly, for a 30 in. set, we have $(3h)^2 + (4h)^2 = 30^2$. $h = 6$. The dimensions of a 30″ set are 18 by 24 for an area of 432 sq. in. The percent increase is $\dfrac{432 - 192}{192} \times 100\%$ or 125%.

25. A general solution to this problem is: Suppose that the circumference of the earth is C. Then the diameter, d, is C/π. Now, if the circumference is increased by 1 yard, then the diameter, D, is $(C + 1)/\pi = C/\pi + 1\ \text{yd}/\pi = d + 12$ in. So things about a foot high could crawl under the rope. These might include an ant and a small dog. A large dog, pig, and cow are doubtful.

27. The house plan will vary among students.

SECTION 13.1

1–5.   A quantity is a constant if its value does not change and a variable if it assumes different values.

1. Because the path of the earth is elliptical, not circular, the distance from the earth to the sun is a **variable**.

3. Because the number of days in a year has 2 values, 365 or 366 in leap year, it is a **variable**.

5. The time for a ball to roll down a ramp is a **variable** because it depends upon the beginning position of the ball on the ramp.

7. The variables for this taxi company are the distance driven measured in fifths of a mile, x, and the total fare, f.

9. $F = 0.75 + (5 \text{ mi})(5 \text{ fifths/mi})(0.2) = \$5.75$

11. The salesman receives a salary of \$5000 and a commission of 0.01% on sales. The commission of 0.01% on dollar sales is the same as a commission of 0.0001 on dollar sales or a commission of 0.1 on sales in multiples of thousands. So, for s in thousands of dollars, **T is given by 5000 + 0.1s**.

13. The formula is: **B2: 0.1 ∗ A2 + 5000**.

15. $18 - (0/2) = 18 - 0 = \mathbf{18}$. $18 - (2/2) = 18 - 1 = \mathbf{17}$. $18 - (-8/2) = 18 - (-4) = 18 + 4 = \mathbf{22}$. $18 - (36/2) = 18 - 18 = \mathbf{0}$. $18 - (60/2) = 18 - 30 = \mathbf{-12}$.

17. The length, L, of walkway yet to be paved is **$x - 20$** where x represents the total length of the walkway.

19. Let x represent the price of the items in dollars. Then the cost is **$(x + 0.08x + 2.75)$ or $(1.08x + 2.75)$**.

21. $y = 1.08(125) + 2.75 = \mathbf{\$137.75}$.

23. If R represents the amount of money remaining on the loan, **$R = 3000 - 225x$**.

25. A linear function is a function in which the independent variable is to the first power.
    So $y = -34 + 7x$ is **linear**.

27. Because the independent variable is to the second power, the function is **quadratic**.

29. A quadratic function has the independent variable to the second power. So $y = 3x^2 + 12$ is **quadratic**.

31. $y = ax + 4$

x	1	3	5	10	50	100
y	6	10	14	**24**	**104**	**204**

From the table we see that when $x = 1$, $y = 6$. So $6 = a(1)+4$. Thus $a = 2$ and **$y = 2x + 4$**.

33. $y = ax^2 + 5$

x	2	4	6	10	20	40
y	1	–11	–31	**–95**	**–395**	**–2495**

From the table we see that when $x = 2$, $y = 1$. So $1 = a(2^2)+5$. $1 = 4a + 5$, $a = -1$ and $\mathbf{y = -x^2 + 5}$.

35. Failing to follow rules about order of operations can result in incorrect answers. (The correct answer is 15 because $2 \times x$ should be done first.) To get a solution of 92, Student *A* worked the problem from left to right: $7 + 2 = 9$; $9 \times 12 = 108$; $108 - 12 = 96$; $96 - 4 = 92$. To get a solution of 3, Student *B* solved the problem by subtracting $12 - 12$ first: $7 + 2 \times (12 - 12) - 4 = 7 + 2(0) - 4 = 7 - 4 = 3$.

37. The volume, *V*, of the box is the product of its length, width, and height. So $V = (2x)(x)(x + 5) = 2x^2(x + 5) = \mathbf{2x^3 + 10x^2}$.

39–42. A relationship is linear if equal increments in one variable are accompanied by equal increments in the other variable or if the increments in one variable are proportional to the increments in the other variable.

39. As *x* has equal increments of 1, from 1 to 2 to 3 and so on, the increments of *y* are from –8 to –16, an increase of –8; from –16 to –24, an increase of –8. So the relationship is **linear** and $\mathbf{y = -8x}$.

41. Since for each increase of 3 for the *x* variable there is an increase of 3 for the *y* variable, the relationship is **linear** and $\mathbf{y = x - 2}$.

43. Responses will vary. All students should note that the volume of container *A* is three times the volume of container *B*. Some may continue that, if the containers are cylinders with the same base areas, then *A* has three times the height of *B* or if the heights are the same then the base area of *A* is three times the base area of *B*. Students may consider containers of other shapes. Because the volume of a cylinder is three times that of a cone with the same radius and height, *A* could be a cylinder and *B* a cone of the same height and radius.

45. The first increment on the *x*-axis represents 5 years later, or 1976. The graph directly above that point corresponds to a point on the *y*-axis that is approximately $50.

47. About year 16 = **1987**.

49. The relationship for the Happy Trails To You bicycle shop is proportional because the ratio of the total cost to number of rental days is constant, 9.

51. Answers will vary. Possible stories are given below.

   A bike rental at the Get the Bike Out of Here shop is a flat fee of $4 plus $12 for each day you have the bike. What is the relationship between the number of rental days and the total cost of renting the bike?

   A bike rental at the Happy Trails to You shop is $9 for each day you have the bike. What is the relationship between the number of rental days and the total cost of renting a bike?

53. $N = M + 18$

M	10	24	35	50	52
N	28	42	53	68	70

55. $N = 5M + 5$

M	2	7	10	12	24
N	15	40	55	65	125

57. Draw a few examples, put the data in a table, and look for a pattern. Examine the manner in which you count the seats in the pictures to see if it suggests an algebraic expression.

Number of tables	Number of seats
1	4
2	6
3	8
4	10
...	...
n	2n + 2
24	50

There are a variety of ways to find an expression, and a variety of equivalent expressions can be found. In the example provided, each table can seat 2 people on the outsides and then 2 people at the ends of the tables after they are pushed together. So, the total number of people that can be seated is the number of tables times 2 plus the 2 that sit on the ends. An alternative way to count people is to say that the tables in the "middle" (not counting the end tables) each seat 2 people. If there are n tables, there are $n - 2$ tables in the middle. The end tables each seat 3 people. So, the total number of people that can be seated is $2(n - 2) + 6$. This expression simplifies to $2n - 4 + 6$ or $2n + 2$. To find the number of tables needed to seat 50 people, solve $2n + 2 = 50$; $2n = 48$; $n = 24$.

59. Draw a few examples, put the data in a table, and look for a pattern. Examine the manner in which you count the perimeter in the pictures to see if it suggests an algebraic expression.

Number of triangles	Perimeter
1	10
2	14
3	18
4	22
...	...
n	4n + 6
30	4(30) + 6 = 120 + 6 = 126

Each triangle in the arrangement has its side of length 4 feet exposed. Thus, if there are n triangles, they will contribute $4n$ feet to the perimeter. Each of the end triangles has a side of length 3 feet exposed. These two triangles contribute $2(3)$ or 6 feet to the perimeter. So if there are n triangles in the arrangement, the perimeter is $4n + 6$ feet.

61. Three ways of representing a function are graphs, equations, and tables. The previous 5 problems show functions represented as tables and equations. Graphing these equations on a graphing calculator or plotting data points from the tables will give a graph.

SECTION 13.2

1.

x	0	–3	9	3
$8x^2$	0	72	648	72

The table shows the **–3 and 3** are solutions to $8x^2 = 72$.

3–5.   Responses will vary. Possibilities include:

3. Multiply both sides of the equation by 4.

5. Add $7x$ or 6 to both sides of the equation or subtract 18 from both sides of the equation.

7. Responses vary. One is free to select any value for x or y and solve the equation for the other variable. So:

    For $x^2 + 3x = y$, let $x = 0, 1, 2, 3, 4$ and $y = 0, 4, 10, 18, 28$.

9. $15 – 2x = y$

x	0	1	–1	–5	**9**
y	15	13	17	25	–3

$15 – 2x = –3; –2x = –3 – 15; –2x/–2 = –18/–2; x = 9$

11-5.   All solutions below were found using the properties of equality technique. Some students may choose to use the inspection technique or the try-check-revise technique (with or without a spreadsheet). The answers below are the same regardless of the technique used.

11. $6x – 25 = 35; 6x – 25 + 25 = 35 + 25; 6x = 60; 6x/6 = 60/6; x = \mathbf{10}. 6(10) – 25 = 60 – 25 = 35.$

13. $2.5x + 32 + 1.5x = 20$
    $4x + 32 = 20$
    $4x + 32 – 32 = 20 – 32$
    $4x = –12$
    $x = –3$

15. $\$395 = \$15x + \$12.50; 395 – 12.50 = 15x + 12.50 – 12.50; 382.50 = 15x; 382.50/15 = 15x/15;$ **25.50** $= x. 15(25.50) + 12.50 = 395.$

17–21. Methods of solutions will vary.

17. $x^2 – 17x – 84 = 0 = (x – 21)(x + 4)$; so $x = \mathbf{21, –4}. (21)(21) – (17)21 – 84 = 441 – 357 – 84 = 0$; $16 + 68 – 84 = 0$. Factoring

19. $[-(-13) \pm \sqrt{[(-13)(-13) - 4(-2)(-7)]}] / 2(-2) = [13 \pm \sqrt{113}] / -4 = -5.9075, -0.5925.$
$-2(5.9075)(-5.9075) - 13(-5.9075) = 7.000; -2(-0.5925)(-0.5925) - 13(-0.5925) = 7.000$ Quadratic formula

21. $12x^2 - 11x - 5 = 0; x = [-(-11) \pm \sqrt{[121 - 4(12)(-5)]}] / 24 = [11 \pm \sqrt{(361)}] / 24 = (11 \pm 19) / 24 = $ **5/4, –1/3.**
$12(5/4)(5/4) - 11(5/4) - 5 = 18.75 - 13.75 - 5 = 0; 12(-1/3)(-1/3) - 11(-1/3) - 5 = 4/3 + 11/3 - 5 = 0.$

23–25. The solutions to the quadratic equations in 1 variable are the values of x at which the values are 0. These are the intersection points of the graphs with the x-axis.

23. $x = $ **3, –4.** $(3)(3) + 3 - 12 = 0; (-4)(-4) - 4 - 12 = 0.$

25. Because the graph does not intersect the x-axis, there are no solutions.

27. One approach to solving the equation is guess/check/revise. Suppose $x = -30$. Then $(1/2)(-30 + 12)$ $= -7.5$. This result is small so try $x = -48$: $(1/2)(-48 + 12) = -18$. So the solution is –48. A more systematic approach is to use the properties of equality and inverse operations: multiplying both sides by 2 we have $x + 12 = -36$. Now, subtracting 12 from both sides of the equation we have $x = -48.$

29. The sequence **6, x, 24, +, 13, =** results in **157**.

31. Responses will vary. A possibility is: $2x - 6 = 12$ which can be solved by the key sequence 12, +, 6, =, ÷, 2.

33. Because there are 2 points of intersection, –1 and 7, there are **2** solutions.

35. Maurice took the equation $8 = -4x + 28$ and separated it into two equations: $y = 8$ and $y = -4x + 28$. There is a set of (x, y) values that satisfy the first equation. Similarly, there is a set of (x, y) values that satisfy the equation $y = -4x + 28$. These sets of values are the points on the respective graphs. The point of intersection of the graphs has coordinates that make both equations true. The graphs intersect at $(5, 8)$. So, when $y = 8$, $x = 5$ and $8 = -4(5) + 28$. Therefore, the solution to the original equation is $x = 5.$

37. Responses will vary. Possibilities include: A quadratic equation has 1 solution if the discriminant, $b^2 - 4ac$, is 0. Suppose that $b = 3$, $a = 1$, $c = 9/4$ giving an equation $x^2 + 3x + 9/4 = 0$ with one solution, –3/2. A quadratic has 2 solutions if the discriminant is positive. Suppose $b = 3$, $a = 1$, $c = 2$ giving the equation $x^2 + 3x + 2 = 0$ with the 2 solutions –1, –2.

39. a. $T = 150h + 250$
    b. $1000 = 150h + 250; 750 = 150h; 5 = h$

41. Since area is the product of length and width and since the length is to be twice the width, we have $288 = w(2w)$. So $288 = 2w^2$; $w^2 = 144$; $w = 12$. So the **width is 12 ft and the length is 24 ft**.

43. Without opening the chute, it takes approximately 11 seconds for the diver to free fall to the ground. If the chute takes 5 seconds to open, he has 6 seconds to free fall before opening the chute. However, if the chute is going to have time to slow him down, it needs to be opened almost immediately.

45. a.   Because $49,500 is roughly 3 times as much as $15,750, we can estimate the 1929 value of the stock to be 3 times the value of 175 shares or $189,000.

    b.   In 1932, each share of stock was worth $90. Thus, $49,500 corresponds to 550 shares of stock. In 1929, each share of stock was worth $360, so 550 shares would have been worth $198,000.

47.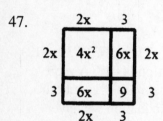

The area of the square $2x + 3$ by $2x + 3$ is:
$4x^2 + 2(6x) + 9 = 4x^2 + 12x + 9$.

49. The solutions of a quadratic equation in one variable are often called zeros because, when solving the equation graphically, the solutions are the values of x for which the y value is zero. Also, when factoring a quadratic in one variable we obtain the solutions by setting the factors separately equal to zero.

SECTION 13.3

1.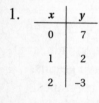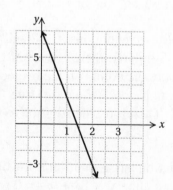

x	y
0	7
1	2
2	–3

3.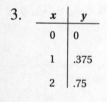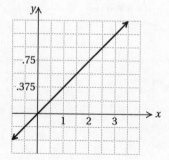

x	y
0	0
1	.375
2	.75

5.

x	y
0	4
1	6
2	8

7. When $x = 0$, $y = 5$. Substituting into the equation produces $y = -4(0) + 5 = 5$.

9. Because the slope is the coefficient of the x term, the slope is **–1**. When the equation is solved for y, the y-intercept is the constant term: **3**.

11. Solving for y we get: $y = -x + 1/2$. The slope is **–1** and the y-intercept is 1/2.

13. The slope is **16** and the y-intercept is **–120**.

15. The data shows that as x changes by 5 units y changes by 40 units. So the **slope is 40/5 = 8**. Since as x **increases** by 5 units y increases by 40 units, if x decreases by 5 units then y decreases by 40 units. So when x goes from 5 to 0 y goes from –80 to –120. Thus the **y-intercept is –120**.

17. The slopes are both –10 and the intercepts are 3 for the first equation and 1/3 for the second. So the lines are parallel.

19–21. Responses will vary. Two lines are perpendicular if the product of their slopes is –1 or if one is vertical and the other horizontal.

19. Since the slope of the graph of $2x + y = -9$ is –2, the equation $y = (1/2)x - 3$ will produce a perpendicular line.

21. Since the slope of $y = -5(3x + 1)$ is –15, $y = (1/15)x$ will give a perpendicular graph.

23–25. Points on a graph are represented as (x, y) pairs and the slope of a linear function can be represented as $m = (y_2 - y_1)/(x_2 - x_1)$ where (x_1, y_1) and (x_2, y_2) are points on the graph.

23. Slope = $(8 - (-1))/(4 - (-2)) = $ **9/6 = 3/2**.

25. Slope = $(8 - 5)/(5 - 8) = -1$.

27.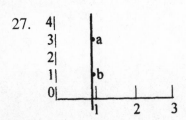

The points a and b on the graph have different y values but the same x values. So as y changes value the corresponding change in x is 0. Thus, were we to calculate the slope we would be dividing by 0, an undefined operation.

29. If a line is to remain in the second and third quadrants only, it must be a vertical line with an equation of the form $x = -b$.

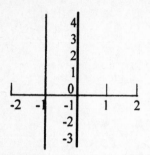

The vertical line $x = -1$ passes through $(-1, 0)$ and $(-1, 4)$ and will remain in the second and third quadrants even if extended.

31. If an equation is of the form $y = mx + b$, m is the slope and b is the y intercept. So: $y = -3x + 3$ has a slope of -3 and a y-intercept of 3.

33. A possibility is $y = 8 - 3x$ which has the same intercepts as $y = 8 - x$.

35. To intersect a line at infinitely many points the second line must be coincident with the first. The 2 lines must have the same slope and y-intercept. One such line is $2y = 24x - 4$ or $y = 12x - 2$.

37. Responses will vary. The horizontal axis might represent time and the vertical axis the number of persons between some individual and entry to an event for which one must show ID. One moves regularly forward until one individual fails to have a validated ID and stalls the line.

39. Let x represent the number of fruit bars, and let y represent the number of cookies. The total cost is represented by $\$0.45x + \$0.34y = \$25.52$. The number of cookies she bought can be expressed as $x + 10 = y$. Substituting $x + 10$ for y in the original equation we get $\$0.45x + \$0.34(x + 10) = \$25.52$; $\$0.45x + \$0.34x + \$3.40 = \$25.52$; $\$0.79x + \$3.40 = \$25.52$; $\$0.79x = \$22.12$; $x = 28$. So the teacher bought 28 fruit bars and 38 cookies.

41. Because the ratio is constant, we can double the total for 150 people and get $2550.

43. The slope of the line is the ratio between gallons of gas used to the cost per gallon of gas. The slope is -200 gallons per 1 dollar. This means that as the price of 1 gallon increases by one dollar, gasoline usage decreases by 200 gallons.

45. Let tu represent the 2-digit number. For example, for the 2-digit number 25, $t = 20$ and $u = 5$. So the number $tu = t + u$ and $(tu)^2 = (t + u)^2 = t^2 + 2tu + u^2 = t^2 + u(2t + u)$ which is the square of tens plus twice the tens plus the units, this multiplied by the units.

Section 13.4

1. AD: $A(-6, 7)$, $D(6, 7)$. $M(x, y)$: $x = (1/2)(-6 + 6) = 0$, $y = (1/2)(7 + 7) = 7$.   **$M(0, 7)$**
    BE: $B(-8, 4)$, $E(-8, -4)$. $M(x, y)$: $x = (1/2)(-8 + -8) = -8$, $y = (1/2)(4 + -4) = 0$.   **$M(-8, 0)$**
    GH: $C(-3, 0)$, $H(6, 0)$. $M(x, y)$: $x = (1/2)(-3 + 6) = 1.5$, $y = (1/2)(0 + 0) = 0$.   **$M(1.5, 0)$**
    FK: $F(0, -4)$, $K(6, -10)$. $M(x, y)$: $x = (1/2)(0 + 6) = 3$, $y = (1/2)(-4 + -10) = -7$. **$M(3, -7)$**

3. $M(x, y)$: $x = (1/2)(-6 + 6) = 0$; $y = (1/2)(0 + 0) = 0$.   **$M(0, 0)$**

5. $M(x, y)$: $x = (1/2)(4 - 5) = -1/2$; $y = (1/2)(3 + -8) = -2\ 1/2$. **$M(-1/2, -2\ 1/2)$**

7–9.   If (x_1, y_1) and (x_2, y_2) are the endpoints of a segment and M is the midpoint with coordinates (m, n), then $m = (x_1 + x_2)/2$ and $n = (y_1 + y_2)/2$.

7. $-2.5 = (-5 + x_2)/2$; $-1 = (3 + y_2)/2$.   Thus $(x_2, y_2) = $ **$(0, -5)$**.

9. $0 = (7 + x_2)/2$; $0 = (7 + y_2)/2$.   Thus $(x_2, y_2) = $ **$(-7, -7)$**.

11–13. Since the distance between 2 points is the length of the segment joining them, if the coordinates of the points are (x_1, y_1) and (x_2, y_2), then the distance is $\sqrt{[(x_2 - x_1)^2 + (y_2 - y_1)^2]}$.

11. $d = \sqrt{[(0-0)^2 + (6-(-5))^2]} = \sqrt{(11)} = $ **11**.

13. $d = \sqrt{[(0-12)^2 + (0-(-4))^2]} = \sqrt{(160)} \approx $ **12.65**.

15–17. If (x_1, y_1) and (x_2, y_2) are two points on a line, the slope, m, is $(y_2 - y_1)/(x_2 - x_1)$.

15. $m = (-11 - 3)/(12 - 5) = $ **-2**.

17. $m = (7 - (-6))/(3 - (-5)) = $ **13/8 = 1.625**.

19–21. The equation of a line with slope m and y-intercept b is $y = mx + b$, so:

19. **$y = 4x - 2$**.

21. **$y = -x$**.

23–25. If a linear equation is solved for y and written in the form $y = mx + b$, then the slope of the graph of the equation is m and the y-intercept of the graph is b. So:

23. Slope $= -3$, y-intercept is -7.

25. Slope is 0, y-intercept is 5.

27–29. If two points of a line have coordinates (x_1, y_1) and (x_2, y_2), then the slope, m, is $(y_2 - y_1)/(x_2 - x_1)$. If the line crosses the y axis at $(0, b)$, then b is the y-intercept and the equation of the line is $y = mx + b$. So:

27. $m = 2/5$, $b = 3$ and $\mathbf{y = (2/5)x + 3}$.

29. $m = 20/2$, $b = 0$ and $\mathbf{y = 10x}$.

31–33. The equation of a circle with center located at (h, k) and with radius of length r is $(x - h)^2 + (y - k)^2 = r^2$.

31. $(x - 0)^2 + (y - 0)^2 = 6^2$. $\mathbf{y^2 + x^2 = 36}$.

33. $\mathbf{(x + 3)^2 + (y - 4)^2 = 4}$.

35–37. Since equation of a circle with center located at (h, k) and with radius of length r is $(x - h)^2 + (y - k)^2 = r^2$:

35. $x^2 + y^2 = 36$ is equivalent to $(x - 0)^2 + (y - 0)^2 = 6^2$. So the **center is at (0,0) and the radius is 6**.

37. The **center is located at (0, 6) and the radius is** $\sqrt{100}$ **= 10**.

39. Responses will vary. But since parallel lines have the same slope, all answers should be of the form $y = -2x + b$, b not equal to -3. So a possibility is $y = -2x + 2$.

41. The product of the slopes of perpendicular lines is -1. The slope of the graph of $-x/2 + 7$ is $-(1/2)$. Since the product of $-(1/2)$ and 2 is -1, the slope of lines perpendicular to the given line is 2 and they have equations of the form $y = 2x + b$. One such line has the equation $y = 2x + 5$.

43. Responses will vary. The equations of graphs of circles centered at the origin have the form $x^2 + y^2 = r^2$. Because the squares of real numbers are positive, the sum $x^2 + y^2$ is positive. Thus the right side of the equation must also be positive and thus cannot be -49.

45. If two lines are perpendicular, neither horizontal, they have slopes with a product of -1. So the slope of a line perpendicular to $y = -x + 2$ is 1. Using the point/slope form for the perpendicular we have $(y - 0)/(x - 2) = 1$ or $\mathbf{y = x - 2}$.

47. The slope of AD is $(4 - 4)/(7 - 2) = 0$. Thus AD is a horizontal line through $y = 4$. The slope of BC is $(8 - 8)/(10 - 5) = 0$. So BC is also a horizontal line. AD and BC are parallel. The length of AD is $(7 - 2)$, 5, and the length of BC is $10 - 5$, also 5. Thus AD and BC are congruent and parallel. Since the slope of AB is not undefined, AB is not vertical. Thus $ABCD$ is a non-rectangular parallelogram.

49. An equation equivalent to $2x^2 + 2y^2 = 36$ is $x^2 + y^2 = 18$ or $(x - 0)^2 + (y - 0)^2 = (\sqrt{18})^2$ So the center is located at **(0, 0)** and the radius is $\sqrt{18}$ or $\mathbf{3\sqrt{2}}$.

51. Point *A* is located at (1, 2) and *B* is located at (8, 8). If in traveling from *A* to *B* the driver maintains the general path from *A* to *B*, that is always traveling in either the +*x* or the +*y* direction, he must travel 6 blocks in the +*x* direction and 7 blocks in the +*y* direction no matter the order in which these blocks are traveled. Thus he must travel 13 blocks total which will take 13 blk(5 min/blk) or about 65 mm.

53. The coordinates of four points can easily be read from the graph: **(0, 4), (3, 1), (3, 7), and (6, 4)**. Since the equation of the circle is $(x - 3)^2 + (y - 4)^2 = (9)^2$, the coordinates of a fifth point may be determined by substituting some *x* value between 0 and 6 into the equation and solving for *y*. For example, for $x = 1$, $y = 4 \pm \sqrt{5}$. So a fifth point is **(1, 6.236)**.

55.

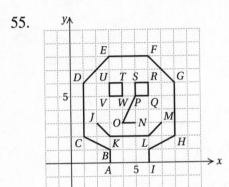

The equations are:

AB: $x = 3$	*LM*: $y = x - 4$
BC: $y = -1/2x + 2\ 1/2$	*NO*: $y = 3$
CD: $x = 1$	*OP*: $y = 2x-5$
DE: $y = x + 5$	*PQ*: $y = 5$
EF: $y = 8$	*QR*: $x = 6$
FG: $y = -x + 14$	*RS*: $y = 6$
GH: $x = 8$	*SP*: $x = 5$
HZ: $y = 1/2x - 2$	*UV*: $x = 3$
ZI: $x = 6$	*VW*: $y = 5$
JK: $y = -x + 5$	*WT*: $x = 4$
KL: $y = 2$	*TU*: $y = 6$

The intersections are:

A: (3, 0)	*L*: (6, 2)	*W*: (4, 5)
B: (3, 1)	*M*: (7, 3)	*Z*: (6, 1)
C: (1, 2)	*N*: (5, 3)	
D: (1, 6)	*O*: (4, 3)	
E: (3, 8)	*P*: (5, 5)	
F: (6, 8)	*Q*: (6, 5)	
G: (8, 6)	*R*: (6, 6)	
H: (8, 2)	*S*: (5, 6)	
I: (6, 0)	*T*: (4, 6)	
J: (2, 3)	*U*: (3, 6)	
K: (3, 2)	*V*: (3, 5)	

57–61. Responses will vary. Possibilities include:

57. $(3, 4) \rightarrow (3 + 2, 4 + 5) \rightarrow (5, 9)$ by translating 2 over and 5 up.

59. $(3, 4) \rightarrow (-4, 3)$ by rotating 90 degrees about the origin.

61. $(3, 4) \rightarrow (3, -4)$ by reflecting over the *x* axis.

63. Responses will vary. Some students may suggest determining all slopes to see if the quadrilateral has 2 pairs of parallel sides, other may suggest that the lengths be determined to see if the quadrilateral has 2 pairs of congruent sides. Still others may suggest a combination of these, one pair of sides, both congruent and parallel, and others may investigate the diagonal relation of mutual bisections.

65.

Diameter	Circle Area	Square side = (8/9) diameter	Square Area
18	254.34	16	256
27	572.26	24	576
36	1017.36	32	1024
45	1589.62	40	1600
54	2289.06	48	2304

area square = 16 x 16 = 256

area circle = 3.14(9)(9) = 254.3

Although the absolute differences between the square and circular area are increasing, the difference as a percent of the circular area appears constant at 0.65%.

CHAPTER 13 REVIEW EXERCISES

1. The number of hours of sunlight in a day is a variable: relatively few in winter, more in summer.

3.

x	0	2	–5	20	–40
$4(2x + 6)$	**24**	**40**	**–16**	**184**	**–296**

5. Let x = the number of hours worked. Then the total cost is $35 + $9.50x$.

7. $C = 35 + 9.50(8) = $**111**.

9. Responses will vary. Substitute for x and solve for y: $(x, y) = $ **(0, –8), (1, –7 3/5), (–1, –8 2/5), (5, –6), (10, –4)**.

11. $3x + 15 = -4x - 20$; $3x + 4x + 15 = -4x + 4x - 20$; $7x + 15 = -20$; $7x + 15 - 15 = -20 - 15$; $7x = -35$; $7x/7 = (-35)/7$; $x = $**–5**.

13. Using factoring, $8x^2 + 12x = 0$; $4x(2x + 3) = 0$; $4x = 0$, $2x + 3 = 0$; $x = $**0, –3/2**.

15. The slope is **–6**, the y intercept is **14**.

17. When x is 3, $y = -4$

19. If (x_1, y_1) and (x_2, y_2) are the endpoints of a segment and M is the midpoint with coordinates (m, n), then $m = (x_1 + x_2)/2$ and $n = (y_1 + y_2)/2$.

$m = (0 + 6)/2 = 3$ and $n = (6 + -9)/2 = -1.5$. So the midpoint is at **(3, –1.5)**.

21. Since the distance between 2 points is the length of the segment joining them, if the coordinates of the points are (x_1, y_1) and (x_2, y_2), then the distance is $\sqrt{[(x_2 - x_1)^2 + (y_2 - y_1)^2]}$.

$d = \sqrt{[(12 - 4)^2 + (6 - 5)^2]} = \sqrt{(65)} \approx $**8.06**.

23. When a linear equation is written in the form $y = mx + b$, m represents the slope of the graph and b the y-intercept.

The graph of $y = 3x - 2$ has a slope of 3 and y-intercept of -2.

25. The equation of a circle with center located at (h, k) and with radius of length r is $(x - h)^2 + (y - k)^2 = r^2$. So: $(x - 4)^2 + (y - (-3))^2 = 3^2$. $\mathbf{(x - 4)^2 + (y + 3)^2 = 9}$.

27. Responses will vary. The line must have a different slope but the same y-intercept. For example: $\mathbf{y = 3x + 11}$.

29. When equations are solved for y, if the independent variable is to the first power, the equation is linear. If it is to the second power, the equation is quadratic. But an equation with the independent variable neither to the first nor to the second power is not linear or quadratic. For example: $\mathbf{y = 2x^3 + 2x - 3}$.

31. The volume of a box is $l \times w \times h$. Boxes like those shown have volumes, V, $(8 - 2x)(10 - 2x)x$.

x	1	1.5	2	2.5	3
V	48	52.5	48	37.5	24

As x increases the volume first increases and then decreases. The volume is a maximum in the vicinity of $x = 1.5$.

33. Responses will vary. Since distinct parallel lines have the same slope but different y-intercepts, the equation of one line parallel to $y = -3x + 9$ is $y = -3x + 2$.

35. Since in a reflection across the x axis $(x, y) \rightarrow (x, -y)$, $(-2, 1) \rightarrow \mathbf{(-2, -1)}$; $(-5, 1) \rightarrow \mathbf{(-5, -1)}$; $(-2, 3) \rightarrow \mathbf{(-2, -3)}$.

37. If 3 and 4 are solutions to a quadratic, then they are the 'zeros' of the equation. So $(x - 3) = 0$ and $(x - 4) = 0$. So the quadratic expression is $(x - 3)(x - 4)$, or $x^2 - 7x + 12$ and the equation is $\mathbf{x^2 - 7x + 12 = 0}$.

39. Responses will vary. The method depends on the information at hand. Essentially there are different ways depending upon what is known. If one knows the slope and y-intercept the equation can be directly written in the form $y = mx + b$, m being the slope and b the y intercept. If the coordinates of two points (x_1, y_1) and (x_2, y_2) are known, then the two point form, $(y - y_1)/(x - x_1) = (y_2 - y_1)/(x_2 - x_1)$ may be used. If the line is on a graph, coordinates may be read from the graph.

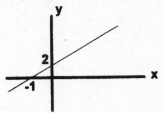